OIL AND INTERNATIONAL RELATIONS

This is a volume in the Arno Press collection

ENERGY
IN
THE AMERICAN ECONOMY

Advisory Editor
Stuart Bruchey

Research Associate
Eleanor Bruchey

See last pages of this volume for a complete list of titles

OIL AND INTERNATIONAL RELATIONS

Energy Trade, Technology and Politics

Michael Fulda

ARNO PRESS
A New York Times Company
New York • 1979

ST. PHILIPS COLLEGE LIBRARY

Publisher's Note: This book has been reproduced from the best available copy.

Editorial Supervision: DIETRICH SNELL

First published in book form 1979 by Arno Press Inc.

Copyright © 1971 by Michael Fulda

Reprinted by permission of Michael Fulda

ENERGY IN THE AMERICAN ECONOMY
ISBN for complete set: 0-405-11957-7
See last pages of this volume for titles.

Publisher's Note: Pages numbered 106, 116, 165, 173 and 277 have been omitted from the reprint edition, due to an error in the pagination of the original text.

Manufactured in the United States of America

Library of Congress Cataloging in Publication Data

Fulda, Michael.
   Oil and international relations.

  (Energy in the American economy)
  Originally presented as the author's thesis, American University, 1970.
   Bibliography: p.
   1. Petroleum industry and trade. 2. Petroleum industry and trade--Technological innovations. 3. World politics. I. Title. II. Series.
HD9560.5.F76 1979   382'.42'282   78-22681
ISBN 0-405-11984-4

OIL AND INTERNATIONAL RELATIONS
ENERGY TRADE, TECHNOLOGY AND POLITICS

by

Michael Fulda

Submitted to the

Faculty of the School of International Service

of the American University

in Partial Fulfillment of

the Requirements for the Degree

of

Doctor of Philosophy

in

International Studies

1970

The American University
Washington, D.C.

© 1971

MICHAEL FULDA

ALL RIGHTS RESERVED

PREFACE

I wish to extend my appreciation to the members of my dissertation committee: Professor Whittle Johnston, Chairman, from the School of International Service, Professor Warren S. Hunsberger from the Department of Economics, and Doctor Sam H. Schurr from Resources for the Future, Inc., for their guidance and assistance in the preparation of this interdisciplinary work. I also wish to thank the Woodrow Wilson National Foundation for the grant of a dissertation fellowship.

TABLE OF CONTENTS

	Page
INTRODUCTION	1
PART I. OIL AS COMMODITY	6
CHAPTER I. OIL TECHNOLOGY	7
Introduction	7
Oil Production	14
Exploration	15
Drilling	16
Offshore	18
Production	23
Conversion	24
Coal	24
Oil Shale	28
Tar Sands	30
Oil Distribution	31
Transportation	32
Refining	45
Storage	62
CHAPTER II. OIL SUPPLY	67
Introduction	67
The Structure Before World War I	84
The Structure Between the World Wars	91
The Present Structure	123
CHAPTER III. OIL DEMAND	146
Introduction	146
Energy Production	151
Primary Energy	151
Organic Energy	152
Inorganic Energy	154
Hydropower	155
Nuclear Energy	159
Fission	159
Fusion	162

Page

Other Energies	170
Geothermal	170
Solar Energy	172
Tidal Power	174
Wind Power	174
Secondary Energy	176
Primary Energy Source	176
Production Costs	181
Energy Consumption	190
Residential Sector	194
Industrial Sector	195
Transportation Sector	198
Motor Vehicles	199
Railroads	203
Vessels	204
Aircraft and Rockets	207
Summary	208
APPENDIX TO CHAPTER III	214
PART II. OIL AS POLITICS	223
INTRODUCTION	224
The Marxist Theory of Imperialism	226
Political Realism	238
Alternative Method	247
CHAPTER IV. OIL AND FOREIGN POLICY	251
Introduction	251
The Oil Boycott	258
Commercial Policy	275
Industrial Oil Importers	275
Exploration and Development	276
Tax Privileges	277
Direct Support to Exploration	278
Other Measures	278
Protection	279
Import Restrictions and Duties	279
Direct and Indirect Subsidies	281
Indirect Protection	282
Research and Development	282
Preventive Measures	284
Developing Oil Importers	286
Developing Oil Exporters	288
The Soviet Bloc	290

	Page

   Foreign Investment Policy 300
     Industrial Oil Importers 301
     Developing Oil Exporters 312
   Foreign Aid Policy 321

CHAPTER V. OIL AND INTERNATIONAL RELATIONS 325

   Introduction 325
   The Pre-World War I International System . . 328
   The Inter-War International System 338
   The Present International System 348
   Toward an Inorganic Energy Society 357

SELECTED BIBLIOGRAPHY 378

LIST OF TABLES

		Page
TABLE I.	THE MAIN OIL-PRODUCING COUNTRIES IN 1913	89
TABLE II.	THE MAIN OIL-PRODUCING COUNTRIES IN 1938	122
TABLE III.	WORLD ENERGY RESERVES AND PRODUCTION	151a
TABLE IV.	THERMAL EFFICIENCY IN FUNCTION OF STEAM TEMPERATURE	183

LIST OF FIGURES

		Page
FIGURE I.	RELATIONSHIP BETWEEN DEGREE OF CONTROL OVER TRANSIT AREA AND TYPE OF TRANSPORTATION SOLUTION	36
FIGURE II.	ENERGY CONVERSION MATRIX	177
FIGURE III.	MAIN ENERGY USE AND SOURCE ACCORDING TO INDUSTRIAL REVOLUTIONS	191
FIGURE A-1.	D-T FUSION POWER PLANT	218
FIGURE A-2.	FUSION REACTOR WITH DIRECT CONVERSION	219
FIGURE A-3.	ENERGY CONVERSION WITH FUSION TORCH D-D AND D-T FUEL CYCLES	220
FIGURE A-4.	FUSION TORCH APPLICATIONS	221
FIGURE A-5.	FUSION TORCH APPLICATIONS	222

INTRODUCTION

The purpose of this dissertation is the analysis of some of the relationships between oil and international relations within a theoretical framework that highlights the salient variables. These two subjects relate in many ways and encompass many variables. Let us begin with the former. Oil is a liquid hydrocarbon found in the crust of our planet and extracted from reservoirs with various characteristics at a certain cost. This introduces the variables of geology and of oil technology and economics. But oil, or a similar refinery feedstock, can also be produced from other fossils, namely from coal, oil shale and tar sands. This technological possibility requires a corresponding increase of the scope of analysis. Seen in another way, and discounting its non-energy uses, oil is a source of energy corresponding to a particular level of energy consumption and production technology. Such a perspective requires in addition the analysis of past, present, and likely future energy competition.

In contrast with coal, which is mainly located in the industrial countries, a large percentage of world oil reserves is located in the underdeveloped world. The increasing demand for oil in industrial countries has caused oil to become the single most important commodity in world trade, both in volume and in monetary value. Outside of the United States and of the Communist world a large part of international oil is produced and distributed by a handful of American and British companies. International oil is a multi-billion dollar industry that affects in varying degrees the balance of payments of oil exporting and importing countries as well as of the home countries of the international oil companies. In this respect oil affects international economic relations. But oil is also a commodity of strategic importance, vital to industry and transport. Lack of oil may paralyze a modern economy and a mobile warfare capability. Whether because of actual or potential oil boycotts in times of peace or of destruction of oil production, transportation and refining facilities in times of war, oil is inevitably enmeshed in international political relations.

How does international oil, with its technological base, fit into the maze of national and international, economic and political relations? Some answers to this question are attempted in this theoretical framework by considering oil both as a commodity in international trade and as an aspect of international politics. The purpose of this somewhat artificial separation is to emphasize some of the relations among the many variables by using levels of abstraction best suited for the analysis of different modes of behavior. Part I, Oil as Commodity, is concerned with technology and economics. Part II, Oil as Politics, attempts to integrate international oil trade with international relations.

Part I consists of three chapters, Chapter I, Oil Technology, is divided into two sections. The first section, Oil Production, examines the past, present, and possible future impact of oil and related technologies on the oil supply of actual or potential oil producing countries. The other section, Oil Distribution, analyzes the economic and political geography of oil transportation, refining and storage, with emphasis on the impact of the international distribution of these facilities on the bargaining power of the various actors involved in international oil trade.

Chapter II, Oil Supply, is a history of the international oil industry. It first examines its peculiar characteristics and why, given existing conditions, international oil trade must be subject to controls outside of normal market mechanisms. It then describes the various historical structures of the international oil industry as defined by the relations between oil companies and governments. These structures span from 1860 to World War I, the Inter-war period, and from World War II to the present. While maintaining some historical continuity the emphasis is put on the period that defined the various structures. Within these periods the emphasis is put on the mechanics of control within the various constellations of interest.

Chapter III, Oil Demand, looks into the future. It attempts to establish how long the significant interaction between oil and international relations will persist. The answer to this question is sought in energy production and consumption competition. It concludes that oil will cease to be a significant factor in international relations after the development of fusion reactors and the transformation of all major types of equipment consuming thermal and mechanical energy to the use of electricity. On the basis of a series of assumptions it establishes a growing demand for oil until the end of the century and a drastic decline in approximately five decades.

Part II, Oil as Politics, consists of an introduction and two chapters. The introduction examines how two opposing theories relate international trade to international relations: the Marxist theory of imperialism with its emphasis on economics and society, and political realism with its emphasis on politics and state. Finding both theories lacking, it opts for a two-pronged approach. The first examines the role of oil in foreign policy, the other analyzes the impact of the historical international systems on the international oil industry.

Chapter IV, Oil and Foreign Policy, focuses on the oil boycott, which lies at the juncture of general foreign policy and of foreign economic policy. It examines how four groups of nations, industrial and developing oil importers, oil exporters and the Soviet bloc, integrate their commercial, foreign investment, and foreign aid policies concerning oil with their general foreign policies. These policies are examined within the present international system.

Chapter V, Oil and International Relations, examines the impact of the national and international, economic and political variables of the historical international systems on the various structures of the international oil industry. After some speculations on the international system of the coming decades that will pave the way towards an inorganic energy society, it concludes that technological obsolescence will divorce the international organic energy trade of the future from international politics.

    A few words on bibliography and methodology. Both reflect the interdisciplinary nature of this dissertation and the unavoidable disciplinary bias of the author. This work draws on three main disciplines: technology, economics and politics. It is written by a political scientist for a degree in international studies. It presupposes that the reader shares his familiarity with politics and his relative ignorance of technology and economics. This results in occasional perfunctory treatment and/or arguments at high level of abstraction for the former, and a possibly excessive documentation and concentration on the elementary aspects of the latter. These are unavoidable perils in the exploration of new horizons.

PART I

OIL AS COMMODITY

CHAPTER I

OIL TECHNOLOGY

This chapter briefly describes the physical process required to transform oil at its source of origin into the products needed at the location of the consumer. It then analyzes how changes in oil production technology and in the distribution of the oil technological process affect national interests.

I. INTRODUCTION

Oil is composed of liquid hydrocarbons. Its chemical composition differentiates it from other organic-bearing sediments by amounts of carbon, hydrogen and oxygen, and by the presence of some impurities such as sulphur and nitrogen. It is a source of energy and a chemical raw material that must be transformed for adequate use. Oil must first be found; this brings us to the vagaries of geology.

> In its simplest terms, the theory of petroleum formation and accumulation has it that oil and gas were formed through deposition and decomposition of small organisms in ancient sea beds. Through subsequent geologic times the resulting hydrocarbon fluids migrated from the source beds to reservoirs in which they were accumulated and trapped. The

history of petroleum exploration is the story of
the search for these trapped accumulations of
oil and gas.[1]

The search for oil is a history of hunches aided by instruments of varying degrees of precision. In some cases oil is close to the ground and leaks to the surface, thereby giving a direct indication of its accumulation. In areas of intensive exploration these reservoirs have by and large been found, thus forcing the prospector to use other methods to guide him in his search. There are two types of oil and gas trapping mechanisms. The structural trap can be identified and located with various instruments that may reveal the possibility of oil and gas accumulation. The stratigraphic trap which is caused by the impermeability of geological layers cannot be detected with present instruments. It is mainly found by chance and by random drilling. Surface oil exploration consists in the use of geophysical and geochemical capabilities in order to find oil and gas trapping mechanisms. Its purpose is to save the costs incurred by random drilling. Down-hole exploration consists in the use of a series of tools to determine the presence of oil in the drilled well and to acquire geological knowledge.

Having by various means established the possibility of an accumulation of oil and gas the next task is to acquire certainty. This can only be achieved by drilling. While the general principles

---

[1] United States Department of the Interior, *United States Petroleum Through 1980* (Washington, D.C.: U.S. Government Printing Office, 1968), p. 39.

of oil drilling are relatively simple, their application to ever
changing geological conditions is a science by itself. The task of
drilling consists of boring a hole in the ground to the desired
depth. Since the obsolescence of cable tool drilling it is done by
rotating a special tool --called bit -- attached to a string of
pipes, either directly with a turbine or indirectly by rotating the
pipes. This serves the dual purpose of cutting in the formation
and of removing the cuttings out of the well, usually by pumping
and circulating water with some additives. The protection of the
well requires the insertion and the cementing of steel casing.

Without going into details, the list of things that can
and does go wrong with this capsule description is impressive.
Suffice it to say that the task inherent in the principle and in
the complications encountered in drilling has produced a vast
petroleum equipment and servicing industry with ramifications in
many sciences and industries. This development has been cost saving
as well as costly.[2]

Leaving geological considerations aside, drilling requires
a concentration of men and materiel on one place at a given time
and is affected by additional factors. According to accessibility
of men and materiel drilling for oil can run the whole gamut from

---

[2]The drilling and equipping cost per onshore well in the
U.S. is around $12,000 for a well of up to 2500 feet, while the
exponential cost of a 15,000 foot well is over $700,000. Some wells
cost up to $2,500,000. Ibid., p. 56.

industrial routine, as in East Texas, to pioneer efforts in some inland jungle. According to surface, drilling may take place on land or in shallow or -- theoretically -- in deep waters. Offshore drilling began in 1946; while the development of its technology is still far from its potential, the present state of the art gives an indication of its possible results. Offshore operations encounter the unique problems associated with the ocean environment. So far, the solutions have taken the form of fixed platforms for drilling rigs with their *ipso facto* depth limitations, and of floating drilling vessels with their vulnerability to weather. Many new concepts are being designed and tested, the ultimate objective apparently being of operating the drilling rig on the ocean floor itself.[3]

As previously stated, the purpose of drilling is to acquire knowledge about the existence of oil and gas accumulations in commercial quantities. If and when the well is deemed productive one must still determine the volume of recoverable oil. This requires further drilling in order to find the horizontal and vertical limits of the reservoir. Once that the volume of recoverable oil has been estimated the wells can then be put into production.

Oil is found in nature under various conditions. According to viscosity, it may range from very light to very heavy; according to formation permeability oil may be recovered in varying percentages of total reservoir capacity; according to reservoir pressure oil may

---

[3] *Ibid.*, p. 47.

or may not flow by gas pressure. "Accordingly, measures aimed at increasing the recovery of oil concern themselves with increasing permeability, reducing oil viscosity, and supplementing or restoring the natural reservoir pressure by the injection of fluids."[4]

The increase in formation permeability is accomplished by fracturing the reservoir rocks with explosives. The use of nuclear fracturing, as tested in the Gasbuggy project, must yet show its potential worth. Oil viscosity is reduced by thermal recovery, the most common methods being steam injection and *in-situ* combustion. The restoration of natural reservoir pressure is accomplished by injecting fluid into the well. According to oil gravity the injected fluid may be natural gas, water or miscible displacement agents.

It should be mentioned at this point that oil can be produced from solid organic-bearing sediments such as coal, oil shale and tar sands. The relative ease in finding and estimating the volume of reserves inherent in the mining of solids is more than offset by high production costs. The production of oil from solid fossils requires conversion technologies competitive with the production costs of natural oil. In the case of coal the conversion process to oil and gas depends on the availability of hydrogen at reasonable costs. The present oil shale conversion process heats the rock and extracts the

---

[4] Ibid., p. 48.

kerogen; its main drawback is that it requires the handling of vast quantities of materiel of which about 90% is waste. Efforts are now being undertaken to retort the shales *in-situ* following conventional or nuclear fracturing. Tar sands are impregnations of hydrocarbons of various viscosity on some rocks, usually on sands. Present technology boils the rocks and extracts the oil; future technology envisages the use of *in-situ* thermal recovery.[5]

Once that oil, regardless of original state, has been found in commercial quantities it must be transported to a refinery, and once refined the oil products must be transported to the consumer. This is done by using containers with independent mobility, by providing mobility of oil in fixed containers, or by a combination of the two. The former are trucks and railroad cars on land, and barges and tankers on water, while the latter are pipelines. Each form of transport has its advantages, depending on volume, type of fluid, distance and surface. Given an adequate volume, pipelines are more efficient on land but must cede to tankers over equal distances in areas of competition. The precise cost efficiency of these forms of transport is difficult to ascertain.[6]

---

[5] *Ibid.*, Chapter 9.

[6] "Many comparisons have been published from time to time on the cost of moving oil in various forms of transport; it is not easy to keep these realistic when costs of one form of transport vary primarily with the utilization of capacity and those of most others mainly with distance." J.E. Hartshorn, *Politics and World Oil Economics* (New York: Praeger, 1967), pp. 70-71.

With a few exceptions oil cannot be used in its original state; it must be transformed into a wide range of specialized products. There are three main types of oil: the paraffinic oil which gives a high yield of gasoline and kerosine, the asphaltic oil with a low yield of light products and a high yield of black oil, and mixed base oils with characteristics in between. There are three main types of end products: the light products, who serve best as energy for explosive combustion, general fuels who yield energy by burning, and lubricants and raw materials for chemical industries.

The refining process consists in the transformation of the various types of crude oils with their respective impurities into a desired range of products. The simplest form is distillation, where the various crude oil components are separated with heat. As is usually the case the crude oil components do not coincide with the desired product mix. This is corrected by various conversion processes such as thermal cracking, polymerization, topping, etc., which permit a wider range of product choices. Finally, whenever necessary, oil is submitted to the process of purification, the most usual being the reduction or elimination of sulphur. All in all, the refining process is capital-intensive, highly technical and requires little labor.

> Any form of technology where labour simply cannot do much of the operating is by definition capital-intensive. An oil refinery is one of the most capital-intensive among a group of industries in which

capitalization is generally high, the 'primary conversion' industries that process basic industrial fuels and materials from their raw state.[7]

Oil does not flow continuously from reservoir to gas station; it must also be stored at certain points, such as ocean terminals, refineries, distributing centers, etc. The cheapest way of storing oil is to leave it in its original reservoir or, at some stage, to pump it into a natural impermeable cavity. Man-made storage comes in expensive containers that benefit from economies of scale.

Let us make a capsule summary of the highlights of oil technology. Finding oil is part science, part art and part chance. The transportation and transformation of oil reflect the essence of the technology of fluids which requires a constant stream of containers. It is a capital-intensive process which benefits from economies of scale, with "diminishing average costs for each extra barrel up to the limits of capacity."[8]

II. OIL PRODUCTION

This section examines some interrelations between oil production technology and national interests. It is assumed that it is in the interest of every nation to maximize competitive domestic oil. Assuming further that every nation with some sedimentary basins has

---

[7]Ibid., p. 74.

[8]Ibid., p. 75.

an actual or potential oil resource basis, advances in the art of exploration and drilling increase the changes of its discovery, while improvements in production techniques increase its economic recovery. In order to simplify this discussion we will arbitrarily classify technological changes as either quantitative or qualitative.

Exploration

"Technology has an impact on the exploration (and drilling) for new oil in three ways: by providing tools and methods for improving efficient discovery; by opening new oil environments (i.e., offshore, greater depths); and by providing cost reduction."[9] Of these three ways only the opening of new environments can be classified as a qualitative change.

The science of exploration has mainly consisted in the improvement and refinement of the basic tool of geophysics. Since 1927, when the seismograph was first introduced, the main success in exploration has been the ability to locate structural traps; its capability has been since enhanced with the use of computers for the interpretation of data. As such, neither improvements in the seismograph nor possible cost reductions in its use has provided for measurable impacts on national interests. In regards to down-hole

---

[9]National Petroleum Council, Impact of New Technology on the U.S. Petroleum Industry, 1946-1965 (Washington, D.C.: National Petroleum Council, 1967), p. 6.

exploration progress has been constant and ingenious. Nevertheless, whatever its present and future capabilities, it shall still require the inherent costly prior act of drilling.[10]

The impact of exploration technology in the new environments being discussed together with drilling technology, there still remains an area for speculation. A possible, though as yet unlikely, qualitative change would be the perfection of a tool capable of locating stratigraphic traps.[11] Similar results could perhaps be achieved by an integrated geological approach.[12] Dealing with an unknown factor speculations are obviously rampant but if one considers the percentage of oil presently derived from these drilled at random traps[13] one could safely conclude that their worldwide discovery would dramatically alter the world supply picture and cause untold impacts on world oil trade.

Drilling

Drilling technology has considerably improved since Drake's first well in 1859. It has progressed from the adaptation of the technology of salt well drilling to an impressive science-based

---

[10]For a compact description in layman's terms of both surface and down-hole oil exploration see: J. Flandrin and J. Chapele, Le Pétrole (Paris: Technip, 1961), part II.

[11]For a discussion of future tool developments see: Impact of New Technology..., op. cit., pp. 80-82.

[12]See A.I. Levorsen, "Big Geology for Big Needs," Bulletin of the American Association of Petroleum Geologists, Vol. 48, No. 2, Feb. 1964, pp. 141-156.

industry. Its principle, however, to drill a hole to a certain depth at a minimum cost, remains the same by definition. Its improvements have consisted in the solution at reasonable cost of the problems involved in drilling in various types of formations, in operating with the factor of increased depth, and in the prevention and remedial actions of the effects of adverse geological conditions. The same is envisaged for the foreseeable future.[14]

One could well classify a cumulative increase in depth as a qualitative change, as increased depth allows for the discovery of deeper reservoirs. Wells over 15,000 feet have been found productive in some countries and either dry or marginal in others, thus providing different benefits from technology, but a meaningful discussion of the above would demand a geological analysis which is neither in the scope of this work nor within the ability of its writer. The main qualitative change that will be discussed is the factor of environment.

It has been previously mentioned that drilling requires a concentration of men and materiel on one place at a given time. The greatest contribution to various national interests has been, perhaps, the ability to produce such concentrations in previously

---

[13] In the U.S. 40%. U.S. Petroleum Through 1980, op. cit., p. 42.

[14] The principle contribution of drilling technology in the future accordingly appears to be in making feasible those operations that formerly could not be done at all, or performed only at unacceptable cost. Ibid., p. viii.

unaccessible land areas. This is discussed later in a different context in view of the fact that the crucial variable in this case is not so much drilling technology as the factor of cost.

Still on land, but requiring a greater input of new technology is a change in environment conditioned by climate. While tropical heat may require some minor adaptations of equipment produced in temporate climates, extreme cold offers a variety of technical challenges. The development of equipment and techniques to cope with the problem of the freezing of liquids favors regions in the Arctic Circle. Such a development, allowing the discovery and production of potentially vast quantities of oil, cannot occur without repercussions on international oil trade.

Offshore. The political implications of advances in drilling technology on land are dwarfed in comparison with the actual and potential political effects of improvements in exploration and drilling technology on or in water. The development of offshore drilling technology has the potential of seriously affecting various national interests by increasing world supply, by changing the pattern of national reserves, and therefore of oil trade and -- perhaps -- by causing various international frictions.

Increased technological capability raises the question of jurisdiction and right of exploitation. For the time being there is only an international agreement to the effect that nations have the exclusive right of exploitation of the natural resources on their respective continental shelves up to at least 200 meters of superjacent waters.[15] There also appears to be an understanding that there are waters over which no nation can exercise jurisdiction.[16] The precise limits are a matter of contention.

The problems related to the jurisdiction over the sea-bed and the ocean floor are numerous and of vast complexity. Among others, they include matters of national security, fishing rights, rights and safety of passage, rights of exploitation of mineral resources, exploitation and conservation of biological resources, pollution, etc. While many of these aspects are interrelated, the exploitation of mineral resources, and specifically of oil and gas, present some special problems.

In considering the exploitation of mineral resources from the sea-bed and ocean floor outside of the agreed minimum of 200 meters, the academic concepts of res nullius and res communis acquire a highly political context. The former allows those capable of

---

[15]Article 1 of the Convention on the Continental Shelf defines it as "the sea-bed and subsoil of the submarine areas adjacent to the coast but outside the area of territorial sea, to a depth of 200 meters or, beyond that limit, to where the depth of the superjacent waters admits of the exploitation of the natural resources of the said areas." American Journal, Vol 52 (1958), as quoted by Charles G. Fenwick, International Law (New York: Appleton-Century-Crofts, 1965) Fourth Edition, p. 448.

[16]It was generally agreed that there is an area of the sea-bed and ocean floor which is not subject to national jurisdiction and that this fact, which seemed obvious, needed emphasizing because of the broad interpretation of which article 1 of the Convention on the

exploiting these resources the liberty to do so without international control and/or profit sharing while the latter does not. It therefore follows that the few nations with the present or immediate future capability of exploiting these resources militate for the freedom for the strong while those nations who cannot opt for the security for the weak.[17]

One can thus see the development of offshore drilling technology creating the following imbalances in national interests. Considering the various interests derived from the minimum agreed upon limits of the continental shelf, a technology which creates the possibility of the development of marine petroleum resources by

---

Continental Shelf was susceptible." United Nations General Assembly, Report of the Ad Hoc Committee to Study the Peaceful Uses of the Sea-Bed and the Ocean Floor Beyond the Limits of National Jurisdiction (New York: United Nations Publications A/7230, 1968), p. 46.

[17] A good example of the high political impact of this question can be seen in the United Nations debates:

The great majority of countries are, for technical, financial or other reasons, not in a position to participate in the exploitation of these resources. The developing and the land-locked countries were specifically mentioned in this respect. In fact only a few highly industrialized countries possess the technical know-how and the investment capital necessary to start any development of these resources. Many delegations stated that this would accentuate the economic imbalance existing between developed and developing countries, and that it would also be an incentive for the former to grab and hold the areas which are most promising.

United Nations Publication A/7239, op. cit., p. 34.

definition eliminates the land-locked nations and those with minimal coastline.[18] On the other hand, it favors such nations with the largest area of continental shelf with sedimentary basins.[19] Going from the possible to the probable, and considering the present state of the arts of exploration and geology, technology will affect promising areas more favorably than others.[20]

Going from theory to practice we find that the minimum agreed upon 200-meter isobath has been extended by unilateral state action and rendered obsolete by technological developments. As of March 1969, and depending on the interpretation, either 19 or 29 coastal states have granted offshore concessions for areas deeper than the 200-meter isobath,[21] the United States having gone as

---

[18] Of the 126 United Nations members 26 are land-locked. National Petroleum Council, *Petroleum Resources Under the Ocean Floor* (Washington, D.C.: National Petroleum Council, 1969), p. 108.

[19] This area can be roughly calculated by multiplying the length of the shore base by the width of the shelf, the latter varying between 1 and 800 miles.

[20] As an illustration, an author has studied the geological potentials of the world continental shelves, rated their oil potential as "excellent," "fair" and "poor," and estimated a potential reserve of 1000 billion barrels of oil from the first two categories. He concluded that about "20% of this area and reserve potential is on the continental shelves of North America." Lewis G. Weeks, "World Offshore Petroleum Resources," *Bulletin of the American Association of Petroleum Geology*, 1965, Vol. 49, pp. 1680-1693, as resumed and quoted in *Impact of New Technology...*, op. cit., p. 80.
Such estimates are obviously vague and disputed, but taken at face value it would show that technological advances would give a proportionately greater benefit to North America due to the fact that 20% of reserve potential is located on less than 20% of the world continental shelves.

[21] *Petroleum Resources Under the Ocean Floor*, op. cit., p. 154.

far as 549 meters.[22] By the same token, recent technological developments have expanded the limits of offshore exploration drilling to 396 meters off the United States coast and possibly to 600 meters in the Caspian Sea.[23] Foreseeable technological advances will erode the 200-meter isobath agreement even further.[24] Barring an international agreement to the contrary it would seem that technology and state practice will replace the outer edge of the geological continental shelf and the 200-meter isobath with the outer edge of the continent.

Insofar as benefits derived from technology in areas of undefined national jurisdiction for the purpose of exploitation of marine mineral resources is concerned, and specially of the ocean floor,[25] the matter is presently in the realm of speculation. It shall undoubtedly benefit the nations with the adequate technology and capital to exploit these resources.[26] The open question is whether

---

[22] Ibid., p. 99.

[23] Ibid., pp. 59 and 154.

[24] Within less than five years, technology will allow drilling and exploitation in water depths up to 1,500 feet (457 meters). Within ten years technical capability to drill and produce in water depths of 4,000-6,000 feet (1,219-1,829 meters) will probably be attained. Ibid., p. 9.

[25] These deep-ocean areas of the world include more than 80 percent of the earth's oceans and 56 percent of the earth's entire surface. By comparison, the submerged portion of the continental masses comprises less than 20 percent of the earth's ocean area. Ibid., p. 115.

[26] Prospects of significant occurences of petroleum under the deep oceans can be neither confirmed nor ruled out. Ibid., p. 8.

23.

other nations will also profit from it, either bilaterally or multilaterally, due to some type of international agreement.

Production

Given a discovered reservoir with a certain volume of recoverable oil under existing technology, any improvement in production technology increases the volume of recoverable oil at a certain cost. Volume increase as well as cost are by and large measurable and thus liable to rational calculations. Given the possible percentage spread between natural and stimulated recovery there is a large incentive to perfect and to use improved production technology. This technology has had a dramatic impact on the oil reserves of the United States.[27] We thus see that improvements in production technology add to the volume of economically recoverable national oil supply. The alternatives are increased costs under existing production technology,[28] costs incurred in domestic drilling, or imports.

The effects of improvements in production technology on national interests depend on net domestic oil postures. For importers with

---

[27]"Largely through these actions, the proportion of economically recoverable oil to the total discovered has been raised from 18 percent to 30 percent over the past 25 years, and much of the Nation's expectations for future domestic oil availability rests upon the outlook for continued progress in recovery efficiency." United States Petroleum Through 1980, op. cit., p. 48.

[28]"There are over 60 billion barrels of oil in known fields that could be recovered in the U.S. if cost were no object." The Oil and Gas Compact Bulletin, Vol. XXV, No. 2, Interstate Compact Commission, Oklahoma City, Oklahoma, 1966, p. 24, as quoted in United States Petroleum Through 1980, op. cit., p. 49.

little or no domestic oil production the impact is neutral from an economic, though not necessarily from a political, standpoint.[29] For net oil exporters it means at least an increased period of revenues, while for nations in between it may make the difference between self-sufficiency and import.

Conversion

The analysis of the effects of improvements in oil production technology derived from the conversion from solid fuels on national interest requires the use of additional factors. The application of conversion technologies on the world distribution of solid fuel reserves does not alone explain their past and present use in some countries as opposed to others. While this phenomenon may be explained in terms of cost advantage, the various cost compositions depend in part on such factors as geographical location, competition with domestic and foreign oil, energy taxation, social costs, etc. Let us examine the situation of every solid fossil

Coal. According to world reserve statistics coal is the second most abundant fossil found in nature. Coal has been partially displaced by oil and gas due to the latters' inherent economic advantages and due to their exclusive use in the internal

---

[29]For example, abundant oil supply in the Western Hemisphere partially due to improved production techniques is in the interest of some West European nations as a counter to potential Middle East oil boycotts.

combustion chamber. It has been seen, however, that this exclusive use can be breached by a conversion from coal to oil and gas at acceptable costs. The past has shown that this conversion process only took place when oil was not available.[30] The question remains open for the future.

The conversion of coal to oil and gas requires the prior act of mining. The cost of coal mining depends on mine characteristics, technology, and on labor and capital costs. Mine characteristics depend on the location and on the horizontal and vertical extension of the coal bed. According to location, coal beds may be at or near the surface, thus allowing for cheaper strip mining, or underground, thereby requiring either shaft, drift or slope mining. According to coal bed extension, mines vary to an extent that allows for production rate differentials from marginal quantities to multimillion annual tons.[31] The efficient application of modern technology to the various mines determines labor productivity.[32] Let us allow for the different

---

[30]Lack of oil was caused either by technological reasons, i.e., coal oil was used prior to the 1850's due to the absence of oil production technology and supply, or by political circumstances, such as the oil blockade of Germany during World War II which forced it to produce expensive synthetic oil. See U.S. Petroleum Through 1980, op. cit., p. 71

[31]In the U.S. the 50 biggest bituminous coal mines produced in 1967 a total of 24.7% of the total U.S. production. National Coal Association, Bituminous Coal Facts, 1968 (Washington, D.C.: National Coal Association, 1968), p. 90.

[32]The U.S. favorable mine conditions coupled with the intensive use of advanced technology and with relatively little government

values of the various types of coals[33] and for the different costs in the sizing and cleaning of coal, assume without further description that various countries have different labor and capital costs, and pursue the analysis of the conversion from coal to oil and gas.

In order to compare the competitiveness between natural and coal-derived oil and gas, one must add to the cost of coal at mine-mouth the costs of conversion and of transportation to the market.[34] One thus arrives at a final cost of which conversion to oil and gas is only a part, albeit a significant one. A price comparison between natural and coal-derived oil and gas must therefore include an analysis of total costs, including transportation costs.

---

inferference contrasts with West European unfavorable mine conditions, inextensive use of advanced mining technology, and government and labor union created featherbedding. The statistics tell the story: output per man per day per net ton of coal is 18.52 in the U.S. and 2.61 in Western Europe. Figures: *Ibid.*, p. 89. The Western Europe figure is an average of the figures for the United Kingdom, Belgium, Netherlands, France and Western Germany.

[33]Anthracite, bituminous, subbituminous, and brown coal and lignite. Not all coal types are equally amenable to hydrogenation or gassification.

[34]Transportation costs depend in part on the location of the coal conversion plant in regards to the mine-mouth and to the market. Given an adequate volume that warrants the laying of an oil or gas pipeline, the nearer the conversion plant to the mine-mouth the greater the savings or the greater costs that would otherwise be incurred by transporting coal to the market either by bulk in conventional transport forms or by slurry pipelines with their processing costs.

It is within this context that one should determine the impact of improvements in coal conversion technology to oil and gas on national interests. Nevertheless, having stated the necessary context of analysis does not imply an in-depth study of the stated variables in view of the limited aim of this discussion. A realistic appraisal of the stated problem would require an analysis of various energy policies and include such factors as coal and oil and gas production and reserves compared to patterns of energy demand, price composition of the respective fuels, possible regional fuel supply shifts, etc. It is assumed that a change in coal conversion technology would have an impact on the international trade of all fuels.

Based on the present pattern of international oil trade, a competitive coal conversion process could have the following impacts on national interests. For countries with little or no production of neither coal nor oil and gas, and assuming no price change in the import of either energy form, the coal conversion process would have no economic impact but might provide some political advantages as a result of fossil exporter diversification. For countries with little or no production of either oil or gas but with actual or potential coal production in excess of present demand, the coal conversion process may cause a change in their respective oil imports. This would be the case of some major oil importers in Western Europe. Finally, for net oil exporting nations, the coal conversion process would cause a net loss of revenues due to some displacement of oil by coal in their major markets.

Oil Shale. Oil shale is the most abundant solid fossil of our planet. It yields kerogen when heated, a refinery feedstock comparable to crude oil. It has been estimated that there is almost 1,000 times as much kerogen as coal in the earth's crust[35] and that a comparable quantity of both fossils could be mined under present conditions.[36] Oil shale is widely distributed among nations with sedimentary basins,[37] and yet in 1966 it was produced in only 6 countries[38] as opposed to 59 for coal.[39] The key to this paradox lies in production economics.

---

[35]United Nations Department of Economic and Social Affairs, Utilization of Oil Shale: Progress and Prospects (New York: United Nations Publication ST/ECA/101), p. 9.

[36]A variable percentage of recoverable coal from a total reserve of 5×10^{12} tons compares to 2.5×10^{12} tons of shale oil in place, accessible under present conditions for oil production. Bituminous Coal Facts, 1968, op. cit., p. 104; and D.C. Duncan and V.E. Swanson, Organic-Rich Shale of the United States and World Areas (Washington, D.C.: United States Geological Survey Circular 523, 1965), p. 9.

[37]Of United Nations members, 53 have reported -- and 26 have possible -- occurences of shale oil deposits. From United Nations Publication ST/ECA/101, op. cit., pp. 32-35.

[38]Brazil, China (Mainland), West Germany, Spain, Sweden, and U.S.S.R. In addition, 5 countries have ceased production. Ibid., pp. 36-37.

[39]Source: U.S. Bureau of Mines.

At a first glance there are certain parallels that can be drawn between coal and oil shale. Both are abundant in nature and mined by similar methods at comparable costs; both can be used to produce electricity or converted to refinery feedstock. Here, however, similarities stop. While there are various qualities of coal and of oil shale, the average coal yields more BTU's that the average oil shale. More important is the fact that coal has an ash content of approximately 10%,[40] while oil shale ranges between 60 to 90%.[41] It is the disposal of this ash that produces the disparity in the production costs of these two fossils, and which explains why the technology of coal production is over two centuries old, while the technology of shale production is still mainly in the future.

Improvements in the oil production technology from oil shale come from two directions. The first comes from improved mining techniques including the disposal of ashes. The finding of commercial use for the ash as well as the extraction of minerals found in the ash are steps in this direction. The second may come from the use of a new technology, i.e., in-situ combustion (fireflooding). Its main advantage is the bypassing of the costly mining and disposal of ash procedures; its relative disadvantage is the loss of minerals that could be extracted from the ash.

---

[40]Percentage for the U.S., United States Petroleum Through 1980, op. cit., p. 74.

[41]Lower range from United Nations Publication ST/ECA/101, op. cit., p. 11. Higher range from U.S. Petroleum Through 1980, op. cit., p. 75.

As in the case of coal, a price comparison between natural liquid and shale oil must include an analysis of total costs, including transportation to the market. As opposed to coal, however, the shale oil industry is by and large in the planning state and has not yet acquired the vested interests and the internal political power of its counterpart.

The determination of the impact of improvements in oil production technology from oil shale on national interests is similar to the discussion on coal conversion technology within the limited scope of this analysis. Nevertheless, there is a long range difference due to the world distribution of oil shale reserves. In the short run, within the order of magnitude of oil shale reserves, their exploitation will benefit those countries with reserves in commercial quantities. In the very long run, however, the large disparity in world oil shale distribution will have a unilateral effect. Whereas 80% of world coal reserves are located in the United States, the Soviet Union and Mainland China,[42] 88% of the world principal oil shale reserves are located in the United States alone.[43]

Tar Sands. The production economics of tar sands are quite similar to oil shale. Tar sands are as a rule more accessible than oil shale but stand equally to gain from improvements in mining and

---

[42]Bituminous Coal Facts, 1968, op. cit., p. 104.

[43]Source: United Nations Publication ST/ECA/101, op. cit., pp. 24-26.

waste disposal technology and from a breakthrough in the techniques of _in-situ_ combustion. The determination of the impact of these improvements on national interests is difficult to ascertain in view of the present sketchy knowledge of world tar sands reserves. It would seem, however, that the distribution of known world tar sands reserves is as asymmetrical as for coal and oil shales. Of the estimated 1,100 billion (1.1×10^{12}) barrels of oil recoverable from world tar sands reserves[44] the lion's share is located in Canada and Madagascar.[45]

III. OIL DISTRIBUTION

The discussion so far has centered on the effects of changes in oil technology on the interests of nations as actual or potential oil producers. In the analysis of the interrelations between technology, geology and national interests it has been seen that the major impact of technological change on national interest is the increase of chances of translating oil resource base into oil reserves. Technology has thus the effect of increasing actual or potential domestic oil production, and of doing this at a reduced cost.

---

[44]Lewis G. Weeks, "Where Will Energy Come From in 2059," The Petroleum Engineer, Vol. 31:2, August 1959, p. A-26.

[45]"The tar sands of Alberta in neighboring Canada have a potential of over 300 billion barrels of recoverable oil -- somewhat greater than the proved reserves so far discovered in the Middle East." United States Petroleum Through 1980, op. cit., p. 71.

The interrelations between oil technology and the interests of nations as oil consumers is more concerned with the distribution of the technological process of oil consumption, the so-called "down-stream" operations. The distribution of this process, namely transportation,[46] refining, and to a certain degree storage and distribution, reflects not only national economic interests, but also such interests based on national calculations of the international situation. A discussion of the "down-stream" operations cannot therefore analyze either interest in isolation of the other.

TRANSPORTATION

It has been previously stated that, given a sufficient volume, pipelines are more efficient than other means of land transportation but that they must cede to tankers over equal distances in areas of possible competition. Let us amplify this statement and analyze some of its economic and political consequences.

> In view of the great variety of conditions determining the competitive merits of tankers and pipelines any comparison should be related to the specific geographical, technical, and economic circumstances of each project, as well as to political considerations.[47]

---

[46] For reasons of convenience, transportation is classified here as "down-stream," at variance with the literature that treats pipelines between oil exporting countries' oil fields to ocean terminals as "up-stream."

[47] Organization for Economic Cooperation and Development, Pipelines and Tankers. A Report on the Effect of the Use of Pipelines on the Transport of Oil by Tankers (Paris: O.E.C.D., 1961), p. 11.

33.

Even while granting that the comparison must be made in each case according to its unique circumstances, let us nevertheless briefly sketch the main relevant factors that enter the calculations. The basic technical characteristics of pipeline and tanker operations are as follows. Pipelines are permanent fixtures which cannot be rerouted. Their oil-carrying capacity depends on their diameter, on the terrain that they cross and on the number of pumping stations. Tankers are by definition more flexible. Their oil-carrying capacity depends on their volume. Theoretically, they benefit from unlimited economies of scale as, ceteris paribus, the perimeter of a container rises with the square of the dimension while its volume rises to the cube.[48] In practice, their size is limited by port facilities and influenced by the depth of relevant waterways.

These technical characteristics have the following effects. Pipeline operations are marked by high fixed charges due to the required heavy initial capital expenditures and by low operating costs. Their profitability is mainly dependent on volume of use up to the limits of capacity. Tanker operations have a much lower fixed charges/operating costs ratio. Their operating costs depend largely on distance. Their profitability depends mainly on tanker volume within the limitations described above. As already stated, tankers

---

[48]Hartshorn, op. cit., p. 65.

34.

are more efficient (ie., have a lower ton/mile cost) than pipelines over _equal_ distances in areas of possible competition. It then follows that, given an adequate volume, pipelines may be _more_ efficient than tankers if they can shorten the latter's route. This brings us to the next variable -- geography.

There are two types of pipelines in regards to tanker competition. The first type is complementary; its only source of competition is a relevant mix composed of barges, trucks or railroads. It transports oil either from the inland oil field to the seaport in producing countries or from the seaport to the inland refinery in consuming countries. The second type "competes with tanker transportation by shortening substantially the alternative sea route when very large quantities of oil are to be transported over a fixed route for a long period."[49]

Before passing to the political analysis one should briefly mention the security characteristics of both means of transportation. The security of a pipeline depends on the degree of control that can be exercized over it. As in the case of all fixed transportation routes such as roads, railroads, rivers and canals, pipeline operations can be disrupted at some point with relative ease, but permanent closure can be accomplished only by lasting physical control. Tankers, on the other hand, are relatively immune to the above when in the open seas

---

[49]O.E.C.D., _Pipelines and Tankers_, op. cit., p. 15.

but vulnerable to opponents with adequate sea and/or air power in case of war. Let us now pass to an analysis of some interrelations between the political and the technical, economic and geographical factors involved in the transportation of oil.

Let us for the time being disregard the question of who decides what oil should go where and assume that in every situation the combination of technical, economic and geographical factors relevant to the transportation of oil has an optimal economic solution which favors some mix of transportation means on its own merits.[50] Let us further assume that there is available capital to put this optimal economic solution into practice. It then follows that every deviation from this optimal economic solution is rooted in political causes. As illustrated in Figure 1 the choice between economic and political solutions is in function of the degree of control over the transit area.

In the case of no transit country, i.e., of internal transportation, we can expect an optimal economic solution in case of effective control over the transit area. In Example A the 4,000 mile Soviet transSiberian pipeline from the Volga-Urals fields to Vladivostock on the Pacific Ocean will, when completed, "replace the present difficult

---

[50]This excludes a situation where, "it may well be that for an integrated oil company, employment of tankers to the full, together with partial pipeline operation, is a more economic proposition than full employment of the pipeline plus laying up the tankers, and this may be particularly true if some of the tankers used are chartered tankers." Ibid., p. 20.

FIGURE I

RELATIONSHIP BETWEEN DEGREE OF CONTROL OVER TRANSIT AREA AND TYPE OF TRANSPORTATION SOLUTION

Transit Countries	Control Over Transit Area	Cause	Solution	Examples	See Text
None	Effective		Economic	TransSiberian pipeline	A
				U.S. Coastal shipping	B
	Ineffective		Political	U.S. World War II pipelines	C
One or more	Effective	Hegemony		'Friendship' pipeline	D
		Common national interests	Economic	Canada-U.S. pipelines	E
				South European pipeline	F
				IPC during mandate area	G
	Ineffective	Conflicting national interests	Mixed	Middle East	H
		Domestic conflicts			
	None	Interdiction	Political	Middle East	I
				Iraq and Jordan to Israel	J
				Mozambique-Rhodesia	K

37.

overland route to the Black Sea and the 10,000 mile tanker route via Suez."[51] In addition to supplying the Siberian market with lower cost oil it will allow the Soviet Union to increase its oil export to Japan and perhaps even to China, depending on improvements in Sino-Soviet relations. In Example B the transit area that affords the optimal economic solution encompasses coastal waters. This solution is used according to the degree of control that can be exercised over the coastal waters. As seen in Example C the United States during World War II built a pipeline system due to the vulnerability of tankers to enemy action. Once that the war was over it resumed the more efficient coastal shipping operations.

In the case of one or more transit countries we can expect an optimal economic solution due to effective control over the transit area caused either by hegemony or common national interests. The case of the hegemonial solution is seen in the 'Friendship' pipeline (Example D) which spans from the Volga-Urals fields to Poland, East Germany, Hungary, and Czechoslovakia, thus replacing the previously used expensive transportation mix. This solution is economic in one sense but hegemonial in another, specially when taken in conjunction with the Soviet oil pricing policy to Eastern Europe enforced by the prohibition of importing significant quantities of cheaper Middle

---

[51]Peter R. Odell, <u>An Economic Geography</u> <u>of</u> <u>Oil</u>, (New York: Praeger, 1963), p. 159.

East oil.[52] This can best be contrasted with Example E, the Canada-U.S. pipelines.

The pipeline between the Canadian Alberta fields and Montreal with branches to the Northern United States could be also, perhaps, classified under the heading of hegemony. While this is true of U.S.-Canadian relations in general it does not apply to this case. The 'Friendship' and the U.S. branches of the Alberta-Montreal pipelines have certain similarities. Both reduce the oil transportation cost in their domestic market due to export volume and serve additional security purposes. On the other hand, while both transport high priced oil, the differential between their high price oil and low cost Middle East oil is borne by the United States in one case and by Eastern Europe in the other. It might thus be argued that the 'Friendship' pipeline is an optimal solution for the Soviet Union only while the Canada-United States pipeline benefits both countries.

The perfect example of effective control over the transit area caused by common national interests is provided by Example F, the South European pipeline. It spans from Genoa to Southern Germany, via Northern Italy, Switzerland and Austria, thereby saving to Switzerland and South Germany the alternative higher tanker freight to the North Sea ports and the barge transportation to the inland refineries.[53] It needs hardly to be emphasized that such a

---

[52] For a more detailed discussion of Soviet oil exports to Eastern Europe see pp. 295-298.

[53] O.E.C.D., Pipelines and Tankers, op. cit., p. 16.

development would have been more difficult to accomplish in any but the present European political atmosphere.

If the South European pipeline is a result, inter alia, of improved political atmosphere, then Example G, the IPC pipeline from Iraq to Syria, Lebanon, Jordan and Israel, is a good reminder of the truism that national interests, and therefore political atmosphere, can clash, thereby lessening the control over the transit area. When the IPC pipelines were laid, British and French common interests in Iraqi oil coincided with the period of their respective mandates over Transjordan and Palestine, and Syria and Lebanon.[54] The pipeline construction through these transit areas was a pure economic solution. In view of the changed political circumstances following the independence of these transit countries one cannot help to wonder whether, with the benefit of hindsight, the IPC pipeline system would not have been routed through Turkey.[55]

The best example of ineffective control over the transit area comprised by one or more transit countries is found in the kaleidoscopical structure of Middle East politics. The four main oil producing states, Iran, Iraq, Kuwait, and Saudi-Arabia are pitted against the main four transit countries, Syria, Lebanon, and Jordan by land and the UAR via the Suez Canal. To the list of producers one should add

---

[54] George Lenczowski, Oil and State in the Middle East (Ithaca, New York: Cornell University Press, 1960), pp. 153-54.

[55] See Ibid., pp. 338-42, for rerouting Irani and Iraqi oil through Turkey to the Mediterranean.

the Soviet Union whose oil exports to Japan are routed via Suez. To the list of potential transit states with mischief-making capability one could possibly add Yemen, South Yemen and Djibouti (presently French), all at the entrance of the Gulf of Aden.

There are three main oil transit areas, the IPC pipelines from Iraq to Syria and Lebanon, the TAP pipeline from Saudi Arabia to Jordan, Syria and Lebanon, and the tanker transit from the Persian Gulf to the Mediterranean via the Gulf of Aden, the Red Sea and the Suez Canal.

The IPC pipeline system has been exposed to politically motivated shutdown and to increased economic liabilities. It is complementary to tankers, but there is a limit past which it ceases to be economic. Regardless of justifications,[56] the limit to the economic rent that can be extracted by the transit countries is set by the cost of laying a new pipeline either through Turkey or to the Persian Gulf (with the latter's increased transportation cost) plus the possible cost of production shutdown during the rerouting period due to transit country retaliation.

The TAP pipeline is 754 miles long. It "avoids the alternative double voyage by tanker of some 3,300 nautical miles each way

---

[56] An example would be the Lebanese assertion that "the geographical position of the countries through which oil passes in transit represents a natural resource for this region exactly as the production of oil does for the producing countries." Ibid., p. 161, quoting Himadeh, "Need to reconsider the Proposal of Revision of the Oil Agreements," An-Nahar and Beirut (Beirut), Nov. 30, 1952.

around the Arabian peninsula and saves the Suez Canal dues."[57]
Calculations similar to the ones used for the IPC pipeline must be
compared with total tanker costs, which depend on tanker size, going
freight rates and on sea route distance, the latter being either the
lane through the Red Sea to the Mediterranean or the one through the
Cape of Good Hope.

The use of the shorter route through the Red Sea to the
Mediterranean includes the following possibilities. The Suez Canal
may be either open or closed. One may or may not create means to
accomodate the present and future supertankers. These means include
either the enlargement of the Suez Canal or the laying of a parallel
pipeline, or the enlargement of the Eilath-Haifa pipeline, from the
Gulf of Aqaba through the Negeb to the Mediterranean. The alternative
uses of the Suez Canal are a purely economic decision while the use
of the Israel pipeline is a highly political act which has been and
will be resisted by the Arab Persian Gulf oil producing states.

We thus see a combination of economic and political solutions
to the oil transportation problem arising from conflicting national
interests. While the transit countries have an interest in the use
of their geographical position in order to further their respective
economic benefits, they have on occasions used the leverage provided
by their position for extraneous political purposes, thus engulfing

---

[57] O.E.C.D., *Pipelines and Tankers*, op. cit., p. 19.

the transportation problem into the quicksand of Middle East politics.

The discussion so far has centered on the control ineffectiveness over the transit area due to governmental action caused by conflicting national interests. It remains to be seen that control may be also hampered or lost due to domestic conflicts that escalate to the use of the <u>ultima ratio</u>. The use of partisan domestic control over the means of oil transportation can be classified according to intended ends. While the means are identical, namely the use or the threat of the use of force, the differences in ends produce different types of conflicts with their respective control possibilities. Within the respective domestic political contexts one may classify these ends as either economic or political.

The first case usually refers to the regional forces aiming for economic benefits derived from the transit of oil through their regions. Their aim is not denial to the central government of oil transit revenues but rather direct or indirect revenue participation. Such would be the case of some Middle East tribes.[58] This type of situation does not overly affect the control over the transit area because its solution lies in the realm of economics and is therefore more amenable to compromise.

---

[58]Lenczowski, <u>op. cit.</u>, pp. 121-25.

In the second case, control over the oil transit areas is much more problematic because political aims are usually of a wider scope than in the former case. These aims may be either regional or national. Regional aims can range from outright secession -- see the effects of the Nigerian civil war on its oil production -- to a larger degree of autonomy, as witnessed by the Kurdish uprisings in Iraq.[59] National aims may be directed against government policies, as shown by the sabotages against Kuwait oil installations following the 1956 Suez Canal intervention,[60] or against the existing government, as sporadically demonstrated by Venezuelan guerrillas.

The majority of these cases have one thing in common, the combination of important political demands with the capability of inflicting a damage disproportionate to their strength. When domestic groups take up arms against the central government in order to redistribute domestic power, the value of oil transit facilities is measured not by the potential benefits that these might accrue to them but by the losses that their denial causes to the government. In this sense the vulnerability of pipelines to harassment or interdiction is ideally suited to pressure national governments. This is especially true for guerrillas who, in order to win, must only show that they have not lost. Even a partial denial of oil transportation facilities can be used as a pungent reminder of their existence.

---

[59]See Harold Lubell, <u>Middle East Oil Crises and Western Europe's Energy Supplies</u> (Baltimore: The Johns Hopkins University Press, 1963), p. 22.

[60]See Lenczowski, <u>op. cit.</u>, p. 338.

The transportation solutions to domestic conflicts resulting in government loss of control over the transit areas depend on individual circumstances. One may, however, indicate the two main variants. If a new situation arises within existing transportation patterns, the intensity of the threat must be weighed against the cost of using an alternative route. Conversely, if a new transportation system must be used within an area of existing domestic conflict, the absence of existing sunken costs in transportation facilities may well indicate the need for the abeyance of production re for the use of a political solution.

The solution to negative control over a transit area caused by interdiction is, by definition, always political. In Example J the IPC pipeline operations from Iraq through Jordan to the Haifa refinery in Israel have been suspended for obvious reasons. As a result, Israel has first relied on Soviet oil and later, after laying the Eilath-Haifa pipeline, on Iranian oil. Example K, the Beira-Salisbury pipeline, is perhaps even more illustrative. A willing transit country -- Portugal -- has been forced to cease pumping oil to Rhodesia due to its loss of control over the relevant transit area -- the Mozambique Channel -- to the British navy backed by a United Nations boycott resolution. The solution to the Rhodesian oil transportation problem has taken the quite uneconomic form of trucking from Mozambique and South Africa.[61]

---

[61] For a more detailed discussion of the Rhodesian oil boycott, see pp. 260-261.

A brief summary of the impact of oil transportation technology and distribution on national interests shows the following salient factors. Cost reducing improvements in oil transportation technology benefit low cost oil producing areas at the expense of higher cost oil areas nearer to the market.

When the distribution of the oil transportation process remains within national borders it benefits that country alone. When it requires the passage through international transit areas it also benefits those who control these areas. Pipelines running through transit countries provide economic and 'political' rents to the transit country governments and, on occasions, domestic groups. Control over tanker routes provides economic and/or political benefits to adjacent coastal states and to those with relevant military power.

Finally, every oil transportation problem has an optimal economic solution with a certain distribution pattern of economic and political benefits. Any deviation from that optimal economic solution relates to the factor of control over the transit area. Its effect is a change in the pattern of benefit distribution.

REFINING

The world distribution of refinery capacity depends on technological, economic, geographical and political factors. While the precise location of refineries depends on individual circumstances one can nevertheless draw some generalizations in order to explain the major trends. This will be done by analyzing this phenomenon on three levels, namely of oil economics, national economics, and international economics and politics. The first level, oil economics, depends on the relationship between refinery economics, transportation and demand. Refinery economics, in turn, depend largely on refinery technology.

It has been previously stated that there are three types of crude oil, paraffinic, asphaltic and mixed based, and three types of end products, namely light products, general fuels and lubricants, and raw materials for chemical industries. Refining technology consists in the transformation processes of the various types of oils into the desired product range.

The first and most important process, distillation, has been compared to the sorting out of pebbles of various sizes by means of a screen.[62] It consists in the separation of the various products in the crude oils according to their respective boiling points. This means that one cannot distill one product, such as gasoline or fuel oil, without producing the others. From this it follows that distillation is a joint process with joint costs.

The second and more expensive process, conversion, is geared to the pattern of product demand. In keeping with the previous allegory, the most common conversion process -- cracking -- compares with the crushing of large pebbles in order to obtain smaller ones.[63]

---

[62] J. Flandrin and J. Chapelle, Le Pétrole, op. cit., p. 312.

[63] Ibid.

It consists in the conversion of some of the distilled products into others which correspond nearer to the market demand. It should be emphasized that one cannot convert all of a crude into only one product. It therefore follows that depending on the range of demand there may be products for which no ready local market is available.

The third and final process, purification, depends on the impurities found in the various crudes and is performed because of either technological, economic or political reasons. Impurities must be eliminated from lubricants for technological reasons, a certain percentage from gasoline for economic reasons, and a larger percentage from gasoline and fuel oils for political reasons, as in the United States antipollution legislations.

Refinery technology affects refinery economics by reducing product costs and by adapting the various crudes to the pattern of product demand. Costs are reduced either by economies of scale or by the engineering of smaller refineries with a lower high fixed cost/low operating cost ratio. The use of the various conversion processes and the engineering of specialized refineries adapts the crude oils to product demand. Having stated the basic facts of refinery economics, let us examine its relationship with transportation and demand, as embodied in the theory of refinery location. According to Tugendhat:

> The theory of refinery location is quite simple. As with all bulk liquids, it is cheaper to transport oil in large quantities than in small bundles, which

means that refineries should always be as close as possible to the place where the products will be sold. At the same time the flow should always be forward, so that refined products do not have to retrace the route down which the crude has already travelled.[64]

While the theory of refinery location may be simple, its application is not, because the crucial variable of demand is both quantitative and qualitative. On one side of the equation one must consider whether the total forward demand supports the economies of scale inherent in profitable refinery economics. On the other side one must consider whether the various products which per force must be produced can find an adequate market. Stated in another way, the location of the refinery should be as near to the consumer as the demand for crude and product allows. It is only where there is sufficient demand both of quantity and of the various products that one can effectuate the savings in the cost differential between the transportation of crude and of products.

Let us now pass to a brief historical review in order to put these theoretical considerations into perspective. "Prior to the Second World War, oil refining was, to a very large degree, located in the major oil-producing areas of the world. At that time, in fact, over two-thirds of the world's refinery was located in the United States, which . . . was then responsible for almost

---

[64]Christopher Tugendhat, <u>Oil: The Biggest Business</u> (New York: G.P. Putnam's Sons, 1968), p. 189.

two-thirds of the total world production."65 This was due to the fact that demand for oil products in the consuming countries warranted neither the costs incurred in refining nor the transportation of crude.

Since the early 1950's there has been a dramatic reversal of the situation. Due to a series of factors, the majority of the world refining capacity is today located in the consuming countries. From an oil economics point of view the causes are as follows. Total demand for oil has increased due to increased energy demand and to the partial substitution of coal by oil, thus allowing for large refineries that benefit from economies of scale. By the same token there has been an increase in the percentage of fuel oil consumption; "the pattern of supplies required thus began to match more closely the most economic pattern of refining the commoner crudes."66 Finally, developments in refining technology, especially in the area of conversion, have helped to adjust the various types of crudes to market needs. The combination of these factors has allowed to benefit from " . . . the greater reductions effected in the costs of shipping crude compared with smaller economies in the costs of moving products . . ."67 which has more than offset the benefits derived from larger economies of scale than are possible by refining the crude in producing countries.

---

[65] Odell, op. cit., p. 109.
[66] Ibid., p. 114.
[67] Ibid., p. 112.

The shift of refinery location from producing to consuming countries cannot be explained by oil economics alone. While oil economics is of predominant concern to the profit maximization units -- the international oil companies -- the location of refineries also affects national interests. This then requires the analysis on the level of national economics.

Refining is performed at a cost which adds to the value of crude oil. The location where this cost is incurred affects the balance of payments. It is therefore in the interests of nations that refining be performed domestically. Insofar as governments are able to pressure the oil companies into locating refineries within their borders they will do so regardless of considerations of oil economics. For analytic as well as historical reasons we can thus classify refineries according to their economic viability.

In the first category national interests coincide with oil economics. This category encompasses the industrial nations and a few underdeveloped nations with sufficient demand for oil and products. The trend towards the location of refineries in consuming nations started in Europe after World War II. The European increased demand for energy was compounded by an increased demand for fuel oil. According to an author, "the great recovery of the Western European economy could not have happened without heavy fuel oil."[68] This

---

[68]Morris A. Adelman, "The World Oil Outlook," in M. Clawson (ed.), Natural Resources and International Development (Baltimore: The Johns Hopkins University Press, 1965), p. 121.

economic rationale was complemented by the desperate hard currency shortage of West European countries who were thus eager to reduce the foreign exchange component derived from the refining process. In order to show the order of magnitude, "it has been estimated that from 1949 to 1955 a saving of foreign exchange of the order of ₤ 160 million was made as a result of the gradual change from product to crude oil imports into the O.E.E.C. countries of Western Europe."[69]

In the second category national interests either ignore or coincide only tangentially with oil economics. This category should be subdivided according to oil importing and oil exporting countries. The first subcategory comprises importer refineries in undeveloped countries which "may be considered uneconomic in that they are manufacturing petroleum products which cost more to produce than the alternative supplies of products refined and shipped from the nearest existing major refinery."[70] Let us see how this process came about.

Since the large scale introduction of low cost Middle East oil on the world market in the 1950's the profits derived from the spread between oil costs and prices has spurned a large competition for crude outlets. This competition has taken place not only among the international oil companies, but it has also been enjoined by newcomers, namely the so-called 'independents' and by the Italian and

---

[69]Odell, op. cit., p. 118, quoting A. Melamid, "Geographical Distribution of Petroleum Refining Capacities," Economic Geography, Vol. 31, No. 2, 1955, p. 168. That is $448 million at the rate of 2.8.

[70]Ibid., pp. 124-25.

Soviet state oil companies. One of the effects of this competition has been the construction of uneconomic refineries in undeveloped countries. After being barred from the U.S. market by import restrictions the 'independents' sought new outlets for their Middle East and Venezuelan crudes in the undeveloped countries and offered refineries in the bargaining process. The Italian state oil company constructed refineries as a separate transaction but in the hope of finding an outlet for its future Middle East crude surplus. The Soviet Union offered oil and refineries on a barter basis. With such competition for their established markets in the undeveloped countries the international oil companies were thus forced to build additional refineries for defensive purposes.

Having stated how this process came about, let us briefly mention who profits from it. The previous arrangements benefited the oil companies only,[71] while the present ones are against their interests. It reduces their total integrated profits, it does not increase their crude outlets while at the same time opening them to greater pressures from the respective governments in view of their increased local investments.[72] These drawbacks, however, have been

---

[71]Discussing the shipping of oil products to undeveloped countries Frankel states that "the traditional and rigid structures of standardized prices for these markets points to the conclusion, however, that the benefits from these economies of scale and of streamlined management accrued to the operating companies rather than to the consumers involved or to their governments." P.H. Frankel, Mattei: Oil and Power Politics (New York: Praeger, 1966), p. 125.

[72]Michael Tanzer, The Political Economy of International Oil and the Underdeveloped Countries (Boston: Beacon Press, 1969), p. 137.

somewhat abated by technological developments.[73] It now remains to be seen what the advantages are to the respective governments. Broadly speaking these fall into the following categories: foreign exchange, crude price, economic development and, on occasions, 'prestige.' The governing factor in the first two categories is control over oil import prices. If a government has the capability to enforce commercial decisions on oil companies it can have the best of both worlds. This, however, is rarely the case, and serious attempts tend to mushroom into international complications.[74]

The realistic alternative is whether the refining capacity should be in the private or in the public sector. If it is in the private sector it implies no expense to the government in the construction of refining capacity but relatively large foreign exchange outlay in the long run due to the higher price of crude oil paid by the affiliates of international companies. Conversely, the construction of refining capacity in the public sector means a large initial capital outlay in both domestic and foreign currency, but a long run economy of foreign exchange in view of the independent

---

[73]"An interesting example of the way in which research and innovation respond to changing circumstances is provided by the changes in refinery economics as political circumstances force the building of smaller refineries. Faced with this demand, the companies quickly developed technology which significantly reduced the cost of smaller scale refining." Edith T. Penrose, The Large International Firm in Developing Countries. The International Petroleum Industry (London: George Allen and Unwin Ltd., 1968), p. 224.

[74]This is specially true if the reduction in domestic and foreign currency is to be achieved by the substitution for cheaper Soviet oil on a barter basis. In Cuba, the result was nationalization

buyer's bargaining power in a buyer's market.75

As any industrial activity, refining is an aid in the development process. This statement, however, should be qualified to the extent that a possible misallocation of human and capital resources may cause unexpected economic and social costs. Finally, there is the intangible matter of prestige where, according to an author, "a refinery has become as essential a part of the paraphernalia of an independent state as an airliner."76 While this may well be the determining factor in some instances one should not discount the other tangible advantages.77

The second subcategory of refinery locations where national interests either ignore or coincide only tangentially with oil economics, refers to those in oil exporting countries. It has been

---

of refineries, in India reduction of crude prices, and in Guinea acceptance by the majors to distribute Soviet refined products.

75For a thorough, if occasionally one-sided, evaluation of the advantages of refineries in undeveloped nations, see Tanzer, op. cit., Chapter 11.

76Tugendhat, op. cit., p. 191.

77An example of a somewhat one-sided evaluation is seen in Odell, op. cit., p. 122, where he contrasts "political and strategic considerations, as well as the impact of national economic policies" of Western Europe and Japan, with "political and economic nationalism . . . in most countries of the world." Actually, in both instances governments act according to their interests, but in one case they coincide with oil company interests while in the other they do not. The difference between both types of national interests becomes then a matter of political semantics.

seen that the profitability of consumer refineries depends largely on qualitative and quantitative product demand. One may thus *a grosso modo* compare the economic advantages of exporter located refineries with the complement of the consumer located refineries. Nevertheless, the absolute -- if not the relative -- growth of refining capacity in exporting nations has been larger, both in number of refineries and in volume of refined oil, than warranted by strict oil economic calculations.

The means by which this has been achieved have ranged from successful bargaining in the concession process, as in the Middle East, to legislation as in Venezuela. The reasons for the successful bargaining of oil exporting nations in view of the world excess supply of low cost oil are dealt with in the next chapter.

The oil exporting nations benefit from domestic refineries. In the first place it satisfies their own domestic energy needs. Secondly it maximizes their oil export revenues, both in domestic and in foreign currencies. Finally, it spurs their industrialization drive by providing the raw material for ancillary industries, specially in the field of petrochemicals.

We have now briefly sketched the national economic reasons that have caused the proliferation of refineries around the globe above the required optimum that could otherwise be achieved on a strict oil economics basis. We can now turn our attention to the third level of analysis, to the economic and political international implications.

Let us review the relevant factors. There are different types of oils which yield different types of product mixes. Light oils yield little fuel oil, heavy oils require the more expensive conversion processes to yield a larger percentage of light and middle end products, high sulphur oil may require expensive purification.

Regardless of refinery location there is an ample movement of products among market regions due to different types of ownership and product demand. The case of a refinery owned by one company processing only one type of crude and selling all of its products in one regional market is an exception. The common occurences are rather of the type where one refinery processes different types of crudes according to source of supply and pattern of product demand. The refinery is either owned by one company which sells its products in various regional markets, or by a few companies who sell their products in one regional market. The third combination involves various companies swapping products between various refineries located in different regional markets. The matching of product demand with available supply within each regional market by the various companies has produced a fine spun web with relatively little flexibility.

Refineries are highly capital intensive installations with high fixed cost/low operating cost ratios, whose profitability depend on maximum utilization up to the limits of capacity. As a result there is little spare refining capacity in the world, thus adding to the international refining rigidity.

In other words the combination of geology, refinery economics and oil company marketing economics provides for a certain rigidity in the qualitative and quantitative supply of products. The corollary is that from the oil company and oil consuming government point of view, the situation would be even worse if the refineries were located predominantly in the producing countries.

> With refineries located at the source of the crude, dislocation involves not only finding an alternative crude supply but also finding somewhere else to refine it. This could in fact be a physical impossibility if the dislocation were large, as refinery capacity in general is not developed greatly in excess of estimated needs.[78]

The implications of the refinery shift from producing to consuming countries are both economic and political. The economic effects have been candidly put forward by the former Secretary General of OPEC. According to Lufti:

> The shift of refinery capacity to the centers of consumption may have some justification on commercial grounds . . . yet it is a dangerous development with regard to the exporting countries of the Middle East. . . . The present trend simply means that . . . these refining facilities can always be used by their owners (the majors to the extent of 61 percent of total capacity) to process crude supplies from other sources to the detriment of the Middle East. In other words,

---

[78]Odell, op. cit., p. 119.

under the present pattern the oil companies concerned can always decide to boost production sharply from an area other than the Middle East if and when such production becomes exportable.[79]

After stating the natural advantages of the Middle East in regards to world refinery location and the unwillingness of the majors to act accordingly, he shows the Venezuelan example of legislated refineries and concludes:

The only way to remedy the situation would be for the host countries and their concessionaires to come to some agreement ot accord on the subject through negotiation, persuasion, or otherwise.[80]

Lufti's 'otherwise' notwithstanding, the shift of refining capacity to the consuming countries is a *fait accompli*. The maximum economic effect of legislated export refineries would be to increase the total operating cost of the oil companies, thus getting nearer to the cost of alternative sources of supply, and of adding to the rigidity of product distribution. Nevertheless, it could neither increase the Middle East bargaining stance nor resharpen the blade of the oil exporter political boycott.

The political effects of export refineries are not limited to their potential denial to consuming nations. They also play a political role due to the magnitude of their cost.[81] In other words

---

[79] Ashraf Lufti, OPEC Oil (Beirut: Middle East Research and Publishing Center, 1968), pp. 18, 19.

[80] Ibid., p. 19 (my emphasis).

[81] Big modern complex refineries may cost up to $200 million. See Jeremy Main, "Meanwhile, Back at the Gas Pump - A Battle for Markets," Fortune, Vol. LXXXIX, No. 7, p. 202.

they are not only a means towards an end but an end in themselves. Export refineries may be located in either producing, transit or entrepôt countries,[82] and subject to control or disruption by governments or domestic groups for economic or political reasons. Effective control is caused either by hegemony or by common national interests between consumer and export refinery nations, coupled with the latter's effective domestic control. Examples of hegemony over export refineries located in producing, transit and entrepôt countries are provided by Britain over Bahrein, Palestine and Aden, respectively. Examples of effective control over export refineries due to common national interests are provided by Venezuelan -- Anglo-Dutch-American relations in producing countries, by Lebanese -- Anglo-French-American relations in transit countries, and by Dutch and British relations with Western Europe in entrepôt countries.[83]

The analysis of the causes of ineffective control over the export refinery is much more involved, because of the larger number

---

[82] In order to minimize political risks oil companies built some export refineries near producing countries in entrepôt areas, usually colonies of company home governments or of friendly consumer nations. Thus, a large proportion of Venezuelan oil is refined in the Dutch West Indies and in Trinidad. This has led Venezuela to legislate that at least 15% of its export oil had to be refined domestically. By the same token, the British built a refinery in Aden after the Iran oil nationalization in order to minimize the dependence on Persian Gulf refineries, notably on the major Abadan refinery. The granting of independence to Aden, presently in South Yemen, has somewhat neutralized these political advantages.

[83] The London and Rotterdam refineries serve the purpose of balancing the various product needs of Western Europe.

of variables. While the common denominator is conflict, the answers to the question of whose conflict with whom, where, how and why, range across the whole strife spectrum, from national economics to international politics. Let us illustrate some of the conflict possibilities:

WHO	1.	Domestic government
	2.	Foreign government
	3.	Domestic groups
WITH WHOM	1.	Export refinery government
	2.	Consuming government/s
	3.	Export refinery owner/s
	4.	Export refinery owner's government/s
WHERE	1.	Producing country
	2.	Transit country
	3.	Entrepôt country
HOW	1.	Disruption
	2.	Destruction
	3.	Confiscation
WHY	1.	Economic gain
	2.	Political denial

It should be clear from the above that the full scale of these possibilities can only bloom in a millieu characterized by civil and foreign wars.[84] Yet the calculation of probabilities depends on

---

[84] The enumeration of these possibilities is not necessarily an academic exercise. This can be seen from an example taken from the oil transportation process, the disruption of the TAP pipeline in June 1969 by Arab guerillas in Israel occupied Syria. It achieves, in the words of its leader, "shock value . . . for the whole Palestine cause." It is aimed at "American interests" and at the Israeli government. The results in regard to the target governments are negligible. It does, however, deprive revenues to the producing country, Saudi Arabia, and to the transit countries, Lebanon and

intangibles and mercurial constellations of interests. Nevertheless, a few conclusions can be drawn.

Export refineries are an economic asset to the respective governments, regardless of their location in respect to the oil transportation process. Depending on previously discussed circumstances they may also benefit oil companies. As one scales the conflict ladder these refineries loose the value of their economic function and become hostages to political ends. In other words, with increased conflict their value is not measured by the economic benefits that they distribute but by the redistribution of power that their denial may cause.[85] This is as true for oil companies, whose vulnerability lies in their large investment, as for governments vis-à-vis domestic groups and foreign governments.

Syria. It forces the Aramco parent companies to reroute the oil via the Cape of Good Hope. The transport cost differential may or may not be passed on to the European consumer. Net benefits: prestige for the guerillas and more business for tankers. It takes little imagination to construct as bizarre a scenario for refineries. See the interview with Dr. George Habash, leader of the Popular Front for the Liberation of Palestine (PFLP), in Time, June 13, 1969, p. 42.

[85]This oversimplified statement describes a mixed bag of conflict representing various participants playing a combination of zero-sum and non-zero-sum games for different stakes according to undefined rules. Generally speaking, when the game, regardless of type, is played for economic stakes it tends to be symmetrical, measurable, and within a predictable outcome range. On the other hand, when played for political stakes with the use or the threat of the use of force, it becomes asymmetrical, unmeasurable and with unpredictable outcomes. Games of either type played for mixed stakes fall between both categories of characteristics and outcomes.

A capsule resume of the impact of the world distribution of refining capacity on national interests includes the following conclusions. Subject to the aspect of control, refineries that convert domestic or imported oil benefit the respective nations by reducing their foreign exchange outlays, by providing additional economic and social advantages, and by decreasing their reliance on nations with export refineries. Depending on the constellations of domestic and foreign interests and on the ends and means of the various participants, export refineries may range from economic asset to political liability to the various nations concerned.

STORAGE

As with transportation and refining capacity, the world distribution of storage capacity depends on technological, economic, geographical and political factors. In keeping with the previously used method, the analysis of world storage capacity is done on three levels, namely oil economics, national economics, and international economics and politics. The first level, oil economics, depends on the relationship between transportation economies and demand, and is further influenced by marketing competition and consumer inventory policy. Transportation economics, in turn, depend largely on transportation technology.

The technological rationale for storage is based on the bottlenecks caused by different flow rates among transportation means, and between transportation means and consumer demand. Storage is required in view of the fact that oil flows continuously through pipelines and at intervals with all other transportation means, and because of the different flow intervals between those provided by transportation means and those needed by consumer demand.

The oil economics rationale for storage is based on the variable of volume as determined by the relationship between transportation economics and demand. This is due to the contrast between the low cost inherent in the transportation of bulk and the high cost involved in the distribution of small volumes to the various locations of the final consumers. It is therefore more efficient to transship oil from large to small volume transportation means via storage capacity.

Storage capacity is further influenced by marketing competition and by consumer inventory policy. Due to reasons that are analyzed elsewhere, product distribution is the main area of competition among oil companies. It logically follows that storage capacity for product distribution in excess of market demand is neither needed nor available where private or public monopoly prevails. Consumer inventory policy may warrant storage in excess of immediate demand in order to satisfy fluctuating demand caused by either unexpected or seasonal factors, or to benefit from price advantages offered with the purchase of bulk.

At the second level of analysis we find that oil storage is of no great concern to national economics and that its interests generally coincide with rational oil economics. Oil storage does not generate any significant economic activity apart from its installation and periodic maintenance. This relative indifference changes to great concern at the third level of analysis, at the level of international economics and politics.

Oil storage has two functions: one one hand it is a necessary part of the transportation process, on the other it provides a stock of oil in excess of immediate demand. Both functions may have international effects. It has been seen in the discussion of the transportation process that the main political variable is the degree of control over the transit area. The same reasoning applies to storage. In transit areas under effective control storage is a purely economic necessity; conversely, in transit areas under ineffective control storage may become a political liability, not because of its intrinsic worth -- as in the case of refineries -- but because of its strategic function in the transportation process.

The storage of oil in excess of immediate demand as an instrument of national policy has direct international economic and political implications. For oil exporters this may enhance the capability of exporting at a moment's notice.[86] For oil importers

---

[86] During the 1967 oil boycott the U.S. supplied a significant amount of oil to Western Europe from its spare capacity in Texas and Louisiana. But this spare capacity was not sufficiently geared to rapid increases in production. This can be seen from the bottlenecks that resulted from insufficient transportation and storage capacity, as well as of water and gas handling equipment. See United States Petroleum Through 1980, op. cit., p. 28.

65.

it may mean the availability of oil in an emergency during the time needed to switch to alternative sources of supply.[87] But oil storage is part of a larger system of guarantees against oil stoppages which include "diversification of supply sources, insuring access to spare production capacity and the accumulation of stocks,"[88] as well as the maintenance of a tanker surplus and the stockpiling of alternative means of energy, usually coal.[89]

The storage of oil in excess capacity, regardless of type,[90] for political purposes, is performed at cost[91] which may be borne by

---

[87]"It is necessary, therefore, for any consumer country which is greatly dependent on imported oil supplies to establish and maintain a stockpile big enough to maintain its internal supply at a constant level, or to prevent its reduction below a particular acceptable level, while oil supply sources are being switched." Organization for Economic Cooperation and Development, Oil Today (1964) (Paris: O.E.C.D., 1964), p. 34.

[88]Ibid., p. 34.

[89]For a discussion of the maintenance of surplus of tankers and the stockpiling of energy in the context of protecting Western Europe against Middle East oil crises, see Lubell, op. cit., pp. 136-40 and 146-47.

[90]"Unconventional storage may include such means as underground storage in salt domes or in washed-out salt beds, open quarries or open earthen pits, abandoned mines or railway tunnels, storage in undersea plastic tanks or in tankers as floating storage. Ibid., pp. 141-42.

[91]For a calculation of oil storage costs and their effects on the competitiveness between fuel oil and coal, see Adelman, op. cit., pp. 121-22.

the taxpayer, the shareholder, the consumer, or a combination of the above. Regardless of how the cost is borne, it is an investment in a nonproductive economic activity, and therefore a loss to the national economy paid in the name of security. It seems useful to draw a comparison between the effects of refining and storage capacity in importing countries. Refining capacity benefits the economy and insures quasi-autarky in oil products. Storage capacity in excess of demand is a loss to the national economy but it blunts the effects of the political boycott of crude oil.

In short, we may conclude that the international political effects of oil storage are as follows. Oil storage in exporting countries is an economic asset which may become a political liability in case of loss of control. Oil storage in excess of demand in importing countries is for them an economic loss but a political gain. For exporting countries it may be an economic and political loss or gain, depending on their political relations with their customers.

CHAPTER II

OIL SUPPLY

This chapter examines oil as a primary commodity in international trade, analyzes the peculiar characteristics of the industry, and describes the various historical structures of the international oil industry as defined by the relations between oil companies and governments.

I. INTRODUCTION

Oil is found in reservoirs located in sedimentary basins. The amount of recoverable oil depends on reservoir characteristics such as volume and shape, formation structure and porosity, oil viscosity, gas and water pressure, etc., and on the state of technology at any given time. The wide range of reservoir characteristics accounts for the large difference of quantities of recoverable oil, from marginal quantities in the so-called "stripper-wells" to the billions of barrels in the Burgan field of Kuwait. In order to analyze the supply of oil one must first estimate its potential. This being an arduous task, let us compare the respective procedures used to estimate the potential of oil and of solid fossils.

> Coal, oil shale, and asphalt occur in solid
> bodies that are extensively exposed at the earth's

surface, where their dimensions, character and geologic relations may be observed directly; . . . In short, resources of the solid fossil fuels are at least partially visible and relatively concentrated, whereas the sources of oil and gas are entirely unseen and relatively diffuse.[1]

Once a concentration of solid fossils has been found it is a relatively simple and not too expensive task to determine its volume and content within acceptable limits of precision. By contrast, one cannot determine the presence of oil without drilling; past minimal quantities one cannot estimate the volume of recoverable oil without extensive development. Thus, the cardinal difference in the determination of supply potential between solid fossil fuels and oil is a significant degree of indeterminacy.

The determination of the world oil resource base is an exercise in educated guessing. The history of these guesstimates is a tale of constant upward revisions, of the dynamic relations between better knowledge of geology and advances in technology. Having previously described the latter, let us briefly touch on the former. In view of the woefully inadequate detailed knowledge of world geology the only reasonable procedure is to extrapolate from known data. The most extensively drilled area of the world, both in total footage and in percentage of total area is the United States; but

---

[1]T.A. Hendricks, Resources of Oil, Gas, and Natural-Gas Liquids in the United States and the World (Washington, D.C.: U.S. Department of the Interior, Geological Survey Circular No. 522, 1966), p. 1.

even *its* geology is inadequately sampled. As a result we are faced with extrapolations based on past production of U.S. reservoirs compared with estimates of the world sedimentary basins.

However this might be, the latest estimate of the world oil resource base and of the amount of oil to be ultimately recovered is of 10,000 and 6,200 billions of barrels respectively.[2] If one considers that only a fraction of this self-avowed conservative estimate has been produced in oil's first century, and that the present world production is under 15 billion barrels per year,[3] one can see at a glance that, even with growing production rates the world supply of oil is assured for times to come. Comforting as this knowledge might be, the picture changes somewhat when one considers that in 1966 the estimated proved oil reserves was of the order of 400 billion barrels, of which 250 billion is in the Middle East.[4] But to this later. The point to be emphasized is that if to the considerable resource base of natural liquid oil one adds

---

[2]*Ibid.*, p. 18. This estimate is based on known world geology and on extrapolations from U.S. geology, of which only one-sixth of the total area is considered to be adequately sampled (p. 8). It considers the continental shelves bordering sedimentary areas or basin margins on land to a water depth of 600 feet (p. 16).

[3]The estimated 1970 oil production outside of the Communist countries is under 12 billion barrels. Commission des Communautés Européennes, Tendances Énergétiques Mondiales, Serie Énergie - No. 1 (Brussels: E.E.C., 1968), p. 20.

[4]Tugendhat, *op. cit.*, p. 301.

the enormous reserves of solid fossils capable of being transformed into synthetic oil (of which 2,000,000 billion of barrels is shale oil),[5] one must conclude that the determining factor in oil supply is not the much feared scarcity, but rather demand, cost and price.

Let us first consider the general framework of oil demand as it affects the supply of natural liquid oil. At any given level of energy production and consumption technology the upper limits of oil demand are set by energy competition. Thus the upper price limit of oil used as a general fuel is set by the price of other energy sources delivered at the location of the consumer. By the same token, the present development, however imperfect, of the technology of synthetic oil production now establishes the theoretical upper price limit of presently non-substitutable oil products used in various energy-consuming equipments, notably the internal combustion and thermal expansion engines. Energy competition is the subject of Chapter III.

Energy competition does not operate in a political vacuum. Governments influence oil demand either directly, by compulsion, or indirectly, via costs and prices, by means of actions or inactions directed at general or particular energy consumers and producer groups for economic and political reasons. This influence takes the form of legislation or administrative action in both the private

---

[5]Duncan and Swanson, op. cit., p. 18.

71.

and the public sectors, and may be the result of energy -- or other type policies. This and other aspects are considered in Chapter IV.

Thus, from the above discussion, we will consider the world supply of oil against the background of a large resource base, and within the parameters set by oil demand, whose upper limits are set by energy competition. Within this framework oil supply depends on oil prices whose lower limits are set by production costs. These are presently examined.

The analysis of oil production costs is an exercise in choices of arbitrary variables. Before substantiating this statement, let us briefly review the economics of reservoir production. In keeping with the stated arbitrary vein we will consider the reservoir as the unit of production and disregard the production economics of the single well.[6] It has been previously stated that the amount of recoverable oil depends on reservoir characteristics and on the state of technology at any given time. The combination of reservoir characteristics is responsible for the wide range of production costs, from the billions of barrels at the reputed

---

[6]This choice favors rationality and conservation and fails to consider the production economics of wells under the so-called rule of capture. This refers to competing producers of one reservoir who, in the absence of agreements or regulations own only the oil that they extract from the ground, thus leading to wasteful competition. This is not only limited to the famous East Texas example. Regulations over producers does not guarantee regulation

Middle East figure of 10¢ per barrel to the marginal quantities with production costs higher than market price.

Whatever their characteristics, all reservoirs share the common trait of finite oil resources which is reflected in the production decline curve. There are three ways of computing oil production costs. The first way is to compute the cost of producing one barrel at any given time from existing capacity. This consists of "out-of-pocket" costs. Depending on reservoir characteristics, the incremental barrel of oil may cost more, the same or even less, to produce than the previous barrel. The second way is to compute the cost of a barrel of oil from existing capacity over the whole life span of the reservoir. In that case, "marginal cost is a rising function of output: the greater is the output the higher is the cost of additional output."[7] The third and final way is to compute the cost of a barrel of oil in terms of its replacement from new capacity. In this case one must consider at the very least exploration, drilling, and development costs.

This matter of what could be referred to as inventory policy, i.e., whether oil costs should be computed from the production of existing capacity -- either at any given time or over the whole life span of the reservoir -- or on the basis of replacement costs from

---

among nations -- as in the case of the Madrejones oil field aslant Argentina and Bolivia (I quote from personal experience!) -- or, within nations, among their territorial units. "Even in the Soviet Union there has been slant drilling from one side of a republic's boundary into a pool on the other side." Adelman, The World Oil Outlook, op. cit., p. 73.

[7]Ibid., p. 42.

new capacity, is a subject of intense debate which is discussed later in the text. Suffice at this point to note that while the exact computation of unit barrel cost is an exercise in futility, one can estimate average production costs on a regional basis, and that -- as with all primary commodities -- there are high, low, and in-between production cost areas. Having briefly discussed oil production costs we can now examine their relations to world oil prices.

In order to provide a basis of comparison, let us first examine oil in the framework of primary commodity industries, which "supply foodstuffs and raw materials by agriculture or mining in the form in which they are first exchanged internationally."[8] In contrast to manufactured goods with their wide range of quality differences, primary commodities are of -- if not equal -- then at least of comparable quality. This means from an economic standpoint that price is the main factor which determines the choice between the various primary commodity sources. In conditions of perfect world market competition prices are related to production costs in accord with the supply and demand situation. When a commodity is in surplus price will fall until quantity offered comes to equal quantity demanded; when there is a scarcity, prices will rise until the quantity offered comes to equal the quantity demanded. Price, then, has the function of allocating resources efficiently. In times of surplus, consumers buy more because of lower prices, and producers offer less, withdrawing from production facilities whose use is now too costly. In times of shortage, consumers buy less because of higher prices, and producers offer more, using facilities made profitable

---

[8] J.W.F. Rowe, *Primary Commodities in International Trade* (Cambridge: Cambridge University Press, 1965), p. 2.

by the higher prices.

Most primary commodities deviate in varying degrees from this perfect model. Demand and supply both tend to be inelastic in the short run. Nevertheless, if one were to rank the various primary commodities according to their degree of deviation from the model one would presumably find oil at the end of the spectrum, farthest from perfect competition. Specifically, one would find, in the present conditions of world oversupply and in the absence of a commodity agreement, that high-cost oil is produced in spite of the abundance of low-cost oil, and, adding political insult to economic injury, that further oil is being sought in areas of known high production costs, thus substantially nullifying the resource allocation function of price.

Seven international major companies produce three-quarters of the crude oil in the world outside the U.S. and the Communist countries, own nearly sixty percent of the refining capacity of Europe (the largest consuming center), and probably do even a greater proportion of the marketing.[9] This situation can be referred to as a "partial oligopoly", which means there is a fringe of numerous small independents around the major firms that have oligopolistic market power.[10] The main importance of the oligopoly model is the explanation of the behavior of the major international oil companies who because of their international vertical

---

[9] Penrose, op. cit., p. 88.

[10] Ibid., p. 37.

integration must balance their respective international interests in order to maximize their total integrated profits. This means on one hand that their choice to produce oil from the various areas under their control is made not only on the basis of comparative production costs but also on their investments and markets in other areas. On the other hand, their far-flung international operations also put them under often conflicting government pressures, which result in the dilution of their control.

The above statistics explain the invalidation of the perfect competition model but not the reasons that led to it, nor whether the present situation of control over market forces is due to inherent characteristics in the technology and economics of oil production or rather to historical circumstances. In his Essentials of Petroleum, P.H. Frankel argues that the oil industry cannot operate under competitive conditions.[11] Rather than to describe his theory in detail we will sketch its essential points in order to show its causal links.

> The argument can be summarized as follows: because of the uncertain results of exploration, the high overhead costs at all stages of the industry, and a high inelasticity of demand in the short run, the industry is not 'self-adjusting' in the sense that a fall in prices significantly chokes off supply or stimulates demand. . . . For

---

[11] P.H. Frankel, Essentials of Petroleum: A Key to Oil Economics (London: Chapman and Hall, 1940).

> these reasons, the industry is subject to continuous crises in the absence of reasonable control over supply. . . . and '. . . for technical reasons alone the formation of paramount oil concerns was inevitable; . . .'[12]

We have thus the first link, namely between oil technology and economics, and the size of the oil companies. The next link to be established is between ends and means.

> Such a role could only be fulfilled by the large firms if an important bottleneck in the industry could be brought under their control. For Standard Oil in the U.S. in the last decades of the nineteenth century, the relevant 'bottleneck' was transportation, but for the international industry the key proved to be crude-oil reserves.[13]

Having established the link between the necessity of control and the means used to obtain it we finally arrive at the position of the necessity of vertical integration and of noncompetitive behavior.

> Vertical integration is held to be necessary because efficient operation requires a continuous and secure flow of supplies, and big firms with large and far-flung markets and heavy investments cannot afford to be dependent on others for their supplies unless, of course, they are securely tied up under long-term contracts. Hence !. . . whatever we may do, the fundamental factors come to the surface; the oil industry to exist at all, calls for concerted effort and, however often a cooperative structure may have been disturbed or broken up, it will soon begin to form again.'[14]

---

[12]Penrose, op. cit., pp. 165-66. Quotation from Frankel, op. cit., p. 85.

[13]Ibid., p. 166.

[14]Ibid. Quotation from Frankel, op. cit., p. 97.

Let us briefly examine the firmness of the links between oil technology and the present structure of the international oil industry. This theory is in effect a combination of technological analysis and historical interpretation which, as any such combination of variables, requires various standards of objectivity and empirical proof. The technological analysis rests on two foundations: economies of scale leading to a natural monopoly and strains in adjustment due to the uncertainties of exploration. We can neglect the "downstream" economies of scale as this would require an *a priori* acceptance of the need for vertical integration.

If we revert to the previous discussion on oil production costs we find that they can be computed either on the basis of existing capacity -- at any given time or over the whole life span of the reservoir -- or on the basis of replacement costs from new capacity. The basis of the computation determines, within the framework of reservoir characteristics, whether production costs per additional barrel are equal, less, or more than the previous one. A natural monopoly requires economies of scale with progressively decreasing costs per additional unit of output.

On the basis of the early history of the United States oil industry and on present Middle East reserves, Frankel sees the industry norm as one of decreasing costs. Adelman challenges this assertion, dismisses the legal milieu of the United States as irrelevant to the theory, and introduces into the analysis the factors of long-range time and demand. According to him the relevant period of time is not the ultra-short initial production phase but the whole life span of the reservoir; thus, any additional production is done at increasing costs.[15] Shifting his level from the economics of the firm to the world industry he notes in addition that in view of rapidly increasing demand that the realistic production cost is the replacement cost from new capacity.[16]

---

[16]". . . petroleum production cannot be and is not geared to the ultra-short run; it is too expensive. Cost are considered over the remaining life of the field, which at the beginning is the whole field, and decisions on investment and production are made on that basis. Over the remaining lifetime of a field or well, marginal cost is a rising function of output; the greater is the output the higher is the cost of additional output. Even neglecting the cost of well workovers, fracturing, waterflooding, etc., the basic fact about oil and gas production is the petroleum decline curve." Adelman, op. cit., p. 42, italics in the original.

[15]". . . it is true that once producing capacity is in operation, the well dug and equipped, operating costs are very low. But this fact has no importance in an industry where capacity must be increased 10-15 percent annually because of growing consumption and natural loss of pressure and capacity. To sell crude oil today at a price which would not cover the cost of replacing capacity when one is actually engaged in digging and equipping new wells is unreasonable and is not done . . . Therefore the relevant incremental cost is not the very low operating cost but the sum of development plus operating cost." Adelman, Oil Prices in the Long Run (1963-75), p. 158, as quoted in Penrose, op. cit., p. 168. Italics in the original.

A juxtaposition of these opposing views shows the difficulty of comparing theories using different variables. On strict technological grounds Adelman's theory is hardly debatable, but one hastens to add that their technological disagreement does not extend to their respective analysis of very large oilfields, and that their differences are more of degree than of kind.[17] In Adelman's words:

> There is always a basic qualification in any kind of scientific disagreement. Within certain limits, two theories may be equally good in explaining some fact of nature or of society; past those limits, one may break down and the other remain. But if we have no desire to go past these limits, we do not need to choose.[18]

---

[17]We can see this convergence in views from the following quotations. Adelman's analysis of production costs makes the exception for very large fields:

> "We can at least take note that the best approximation to the cost function of a very large field is a long flat stretch where new capacity can be brought in at about the same cost as the old."
> The World Oil Outlook, op. cit., p. 44.

By the same token, in his discussion on the decreasing costs of oil from the major exporting countries, Frankel makes an exception to this theory by introducing the factor of time:

> "This statement is representative of the actual situation as it now exists in a significant number of producing and potential oil regions. It is bound to apply for the time span of policies now formulated but may lose its validity in a more distant future."

The Relation of World Oil to Developments in the USA. Testimony submitted to the Subcommittee on Antitrust and Monopoly of the U.S. Senate on March 26, 1969, mimeo, p. 10n.

[18]Adelman, The World Oil Outlook, op. cit., p. 46.

Thus encouraged, we will not choose and turn our attention to the other technological factor, namely the strains in adjustment due to the uncertainties of exploration. Let us consider the implications of this factor on the contending theories.

> This 'destabilizing' element plays a central role in Frankel's analysis. But whereas Frankel puts the emphasis on the uncertainties of exploration as the primary cause of alternate glut and scarcity under competitive conditions, Adelman puts the emphasis on the cost of developing discovered reserves as the primary factor preventing the unprofitable flooding of the market that Frankel evidently fears.[19]

If the impact of past discoveries is a matter of interpretation, then the future is obviously a matter of guessing. It has been seen that the potential of future discoveries is of the order of 6,200 billion barrels of oil. The wide open question concerns the reservoir characteristics where these hidden reserves lie. Specifically, whether the as yet undiscovered oil is distributed among "average" reservoirs with "average" production costs, or whether a substantial quantity is located in future "Middle Easts." The answer to this speculation is not academic, as "Such discoveries necessitate substantial readjustments among producers and sources of production which may be unacceptable to governments as well as to industry, and may lead to some form of public or private intervention to mitigate the effects."[20]

---

[19] Penrose, op. cit., p. 168.

[20] Ibid., p. 169.

Summarizing the discussion on the technological analysis on its own merits we find that the link between oil technology and the 'inevitability' of the present structure of the international oil industry is the past discovery of very large low-cost oil fields. On technological grounds alone the theory of natural monopoly as applied to oil seems to be invalid, but an exception must be made for the 'destabilizing' very large low-cost oil fields with their equal or lower cost per additional output over a long period of time. But by introducing the variable of geology we also introduce the history of exploration and discovery with its intrinsic moot interpretations.

It has been stated that the theoretical relation between oil technology and the present structure of the international oil industry is a combination of technological analysis and historical interpretation. While the accuracy of the former is partly blunted by the interpretation of the history of geological discovery the reasoning of the latter is open to serious doubts. This is especially true in regards to the rise of vertical integration in the United States and to its later spread across the world.[21] This will be seen later in the text.

---

[21] For the growth of vertical integration see: de Chazeau, M.G. and Kahn, A.E., Integration and Competition in the Petroleum Industry (New York: Yale University Press, 1959), and McLean, John, The Growth of Integrated Oil Companies (Boston: Harvard University Press, 1954). Also Johnson, Arthur M., "Public Policy and Concentration in the Petroleum Industry, 1870-1911" in Oil's First Century (Cambridge: Harvard Graduate School of Business Administration, 1959).

> The fact that vertical integration and the dominance of very large firms characterize the history of the industry, both in the U.S. and internationally, provides no evidence at all, nor even a presumption, that the industry could not have operated efficiently under a different type of organization; the most we can presume is that either the large firms found it both possible and profitable deliberately to organize themselves and the industry in the way they did, or that the conditions of competition left little effective alternative.[22]

On the basis of such authority we may well reject the "technological theory" of the international oil industry on the grounds of reductionism and excessive determinism. Nevertheless this theory does portray the technological and economic incentives of the various actors that led to the present development of the industry. Thus, even a partial acceptance of this theory with its emphasis on the need for control brings us down from the lofty spheres of allocation efficiency of world resources to the pragmatic arena where various actors maximize their respective interests within the constraints imposed by their operational milieu -- the international oil industry structure.

The concept of structure as applied to the study of patterns of relationships is a heuristic device whose usefulness

---

[22]Penrose, op. cit., p. 167.
To which Frankel concedes that: "If the oil industry had not been so concentrated and integrated before the 'new' areas of production came in during the last forty years or so, it is conceivable that the channels of refining and distribution would have become sufficiently diversified for a market system in the classical sense to develop, fit to take care of the production available. . . . This is now an idle thought in the circumstances prevailing; . . ."
Mattei: Oil and Power Policies, pp. 103-104.

is its ability to order any number of variables according to some desired pattern. If, with some reservations, one accepts the need for control as a given we will define the structure of the international oil industry by the number and type of actors, their economic and political power, and their stakes of conflict. As emphasis is stressed on change we will thus discern various structures based on abstractions of historical experience. This then requires a narrative of the history of the industry, but it will be done on a selective basis. While maintaining some historical continuity the emphasis is stressed on the period that defined the various structures. Within these various periods the analysis will stress the mechanics of control within the various constellations of interests.

According to the stated variables we can distinguish three historical structures of the international oil industry: from 1860 to World War I, from World War I to World War II, and from World War II to the present. These periods coincide or largely overlap with the major historical international systems of the last hundred years. The relationship between international oil industry structure and international system is by no means direct, mainly due to the indpendent variable of geology which gives a random effect to the choice of actors. Nevertheless, the incorporation of the structures into the systems will be attempted in the last chapter. In the meantime let us turn to the beginning of the industry.

II. THE STRUCTURE BEFORE WORLD WAR I

The main characteristic of this period is of a few major companies competing mainly over markets for oil produced from few countries. Barely three years after Drake's first well oil became international and acquired some of the essential characteristics of the present industry: energy substitution, international trade, tariffs, and even assistance to a war effort! ". . . by the end of 1861 kerosine made from oil had entirely displaced the coal variety."[23] In the same year, in search of a less competitive market than the one prevailing in the United States with its odd mineral laws,[24] oil was made to cross the Atlantic,[25] and was promptly met by its first tariff.[26]

---

[23]Tugendhat, op. cit., p. 11.

[24]In the early 1950's, i.e., before present government intervention and referring to the 1860-1865 period, Gerretson wrote:

> "The system, still characteristic today of the American petroleum trade, of stormy large quantities of oil, brought up at a low price during periods of overproduction, as a basis for an export business, dates back in its general features to these years."

Frederick Carl Garretson, History of the Royal Dutch, Vol. I (Leiden: E.J. Brill, 1953), p. 25.

[25]1861: "November - the first shipload of petroleum to cross the Atlantic is the 224-ton sailing ship Elizabeth Watts. It is loaded at Philadelphia with 3,000 barrels of oil headed for London . . . and marks the beginning of regular traffic in full-cargo shipment of oil. The shipping rate is one dollar a barrel plus a 5 per cent commission to the owners of the vessel." James A. Clark, The Chronological History of the Petroleum and Natural Gas Industries (Houston: Clark Book Co., 1963), p. 31. Italics deleted.

It also began to make an important contribution to the war effort of the Northern States in the Civil war, since they desperately needed a new source of foreign exchange to compensate for the loss of the South's cotton.[27]

"Kerosene was the most important of the industry's products until the second decade of the twentieth century. In the 1860s more than half of the kerosene refined in the US was exported; by the 1880s petroleum and its products, most of which went to Europe, ranked fourth in the value of exports from the US."[28] This export was continued in spite of the beginning doubts about the magnitude of reserves.[29]

Between 1873 and 1882 Rockefeller and his group formed alliances with other oil men and the great Standard Oil Trust was created which obtained extensive monopolistic control over transportation and refining in the US. Not only did Standard Oil dominate the domestic industry for a long time but it reached out to the markets of Europe and the Orient, quickly becoming the world's most important supplier of oil products.[30]

---

[26]1862: "January - British government places a one cent per gallon tariff on petroleum imported from the U.S." Ibid.

[27]Tugendhat, op. cit., p. 11.

[28]Penrose, op. cit., p. 54.

[29]1874: "October - Fifteen years after the first oil well was drilled, Pennsylvania's State geologist Wrigley estimates that the United States has enough petroleum to keep its kerosene lamps burning for only four years, becoming the first doom forecaster." Clark, op. cit., p. 49.

[30]Penrose, op. cit., p. 53.

In regards to Standard Oil's competitive methods in areas where it did not enjoy virtual monopoly due to its stranglehold on refining and transportation, "The most commonly employed device was selective price cutting financed by over-charging in other places where Standard enjoyed a monopoly."[31]

> In Germany, Galicia, and Rumania there had been a certain amount of crude oil production for some years, and in Britain, France, and Germany kerosene was also manufactured from shale, lignite, and coal . . . None the less, nothing could have stood in the way of American oil, and Standard would have dominated Europe as thoroughly as the United States if it had not been for the arrival of Russian oil.[32]

For centuries oil had been known to exist in the Baku area. It had provided ". . . the fuel for the sacred and eternal fires of the Zoroastrians . . . but the commercial possibilities of the deposits were not grasped by the Tsarist regime until 1873, when it was decided to allow private prospectors in the area."[33] The Nobel brothers tried via transportation monopoly to imitate the practices of Standard Oil but were thwarted by the Rothschilds who provided the capital for the construction of the Baku-Batum railway, thus connecting the land-locked Caspian Sea to the Black Sea, and thus to the Mediterranean. "With the completion of the railway in 1883 the growth of the Russian oil industry was incredible . . ."[34]

---

[31]Tugendhat, op. cit., p. 31.
[32]Ibid., pp. 35-36.
[33]Ibid., p. 34.
[34]Ibid., p. 36.

Europe became a kaleidoscope of cartels and
price wars as Standard, the Nobels, the Rothschilds,
various Russian independents, and the occasional
American independent jostled for position. In
general, they did not sell direct to the public,
but through wholesale distributors and retailers.
Simultaneously in some countries there would be
all-out price cutting, while in others there would
be a division of the market between a number of
alliances, and in a third cartels including all the
main competitors might be formed. The pattern was
continually changing, and companies that were allied
in one place could be at war in another.[35]

A further blow to Standard's monopoly was delivered by Marcus Samuel, the founder of Shell, with his famous 1892 coup which involved the construction of safe tankers which could pass through the Suez Canal, thus opening the road for Russian oil to capture Asian markets. At this time still another actor came on the stage.

In the Far East a small Indonesian company was
chartered in 1890 to exploit the oil of the Dutch
East Indies, and in 1900 a Dutchman with the commer-
cial aggressiveness and brilliance to match that of
Rockefeller himself became its manager -- Henri
Deterding.[36]

Deterding brought together the rivals of Standard into a single marketing organization to cater to the Asian market from oil produced in Russia, the Dutch East Indies and Burma. Between the years 1898 and 1901 Russia overtook the United States in oil production. This was due on one hand to the depletion of the

---

[35]Ibid.

[36]Penrose, op. cit., p. 55.

Pennsylvania oil fields and on the other to the active export encouragement policy of the Russian government.[37] "By the end of the century there was nowhere left where Standard could be sure of earning high profits to offset any losses that might occur in the United States should new rivals appear on the scene."[38] In 1901 the Spindletop oil well errupted, thus giving birth to the Texan oil industry. This new source of ample, cheap crude oil proved to be the foundation of new integrated giants: Gulf and Texaco. "That event, a turning point in our history, marked the end of the oil monopoly and the beginning of the liquid fuel age."[39]

Soon after the discoveries in Texas and Oklahoma further oil was found in Mexico, thus closing the roster of countries that were producing oil at the outbreak of World War I. Their respective production in 1913 is shown in Table I.

The essence of the pre-World War I structure can be captured in the rivalry between Standard Oil and Shell.

---

[37]See A. A. Foursenko, *Neftianie Tresti i Mirovaia Politika. 1880-e godi - 1918g.* (Moscow: Nauka, 1965), p. 191.

[38]Tugendhat, op. cit., p. 36.

[39]John Bainbridge, *The Super Americans* (Doubleday), as quoted in Tugendhat, op. cit., p. 37.

TABLE I

The Main Oil-Producing Countries in 1913

Country	Millions of Barrels
United States	237.6
Russia	62
Mexico	27.4
Rumania	13.7
Dutch East Indies	11.5
Burma and India	7.9
Poland	7.9
Total	368.0

Source: Tugendhat, op. cit., p. 72.

Standard had derived its strength from the enormous reserves in the United States which could be shipped to any part of the world. Rockefeller never felt the need to find other sources overseas, and could always deal with a small competitor by flooding his market and cutting prices. Shell, by contrast, had no large home base, and it was Deterding's policy always to secure a source of oil as near as possible to his markets. In this way he was able to cut down transport costs and to build up a chain of fields all over the world, which in the long run was bound to give him far more flexibility than his American rivals.[40]

In regards to the international price structure one must differentiate between the one prevailing in the United States and the one in the world at large. Having achieved a control over the

---

[40]Ibid., p. 57.

U.S. market via refinery and transportation, " . . . the old Standard Oil Company had been content to leave the production of crude oil largely in the hands of independent operators."[41] As a result it established a system of "posted" prices, ". . . at which it stood ready to purchase crude at the wellhead, so that numerous transactions occured constantly at published market quotations."[42] This system was continued even after the breakup of the Standard Trust in 1911 and as Southwestern fields came into prominence the U.S. export prices became tied to the prices at Gulf Coast ports. These prices also established world prices in view of the predominance of U.S. exports on the world markets.

This then constituted an international price structure of mixed nature. On one hand there was an open U.S. market in which one, and later several, companies purchased oil according to the supply and demand situation, and where the individual producers were free to accept the purchasing bids according to their respective production costs. On the other hand Standard Oil exported kerosine to the world markets at prices reflecting the competitive situations in the various markets. Standard's sales being worldwide it could

---

[41] Helmut J. Frank, <u>Crude Oil Prices in the Middle East: A Study in Oligopolistic Price Behavior</u>. Praeger Special Studies in International Economics and Development (New York: Praeger, 1967), p. 11.

[42] <u>Ibid.</u>

effectively engage in selective price cutting at home and abroad by recouping its losses in safer markets.

Standard's foreign competitors were left with three alternatives. They could either buy crude at Gulf Coast ports (e.g., Samuel's 1901 contract with Gulf) and attempt to undercut Standard's kerosine prices with no transportation advantages. Conversely they could secure oil on the open market from other sources nearer to their markets at world (i.e., Gulf Coast) prices, thereby achieving some transportation advantage. Finally, as in the case of Shell under Deterding, there was the possibility of securing one's own source of crude, thus integrating backwards, thereby severing the connection with world crude prices. These developments had no great importance on the international industry structure because of the preponderance of US exports, but they set a precedent for the organization of the price structure of the inter-war years.

III. THE STRUCTURE BETWEEN THE WORLD WARS

The structure of the international oil industry during the inter-war period can be characterized as one of intervention by the great Powers during the immediate post-war years for the control over the Middle East oil fields followed by the formation of a cartel by the major oil companies in the wake of large new oil discoveries.

"The war transformed oil from being a source of revenue for tycoons and speculators into a vital industrial and strategic material."[43] As the war became progressively motorized the role of oil became increasingly important; to such a degree that Clemanceau warned Wilson on December 15, 1917: "A failure in the supply of petroleum would cause the immediate paralysis of our armies and might compel us to make a peace unfavorable to the Allies. . ."[44] While Lord Curzon's famous statement that "Truly posterity will say that the Allies floated to victory on a wave of oil" should be taken with a grain of oil-impregnated salt, one should note ". . . Luddendorff's complaint that lack of oil had played an important part in bringing Germany to its knees."[45] Whatever the historical interpretation the fact remains that the end of the war instituted a scramble for what was thought to be the last remaining major source of oil. To be more precise it accentuated a trend which had begun earlier in the century.

Oil became a commodity of strategic importance as soon as various navy authorities realized oil's advantage over coal. In

---

[43] Tugendhat, op. cit., p. 71.

[44] Quoted by Ludwell Denny, We Fight for Oil (New York, 1928), p. 16, and cited in Edwin Lieuwen, Petroleum in Venezuela. A History (New York: Russel & Russel, 1954), p. 18.

[45] Tugendhat, op. cit., p. 113.

contrast to the United States with its ample reserves of oil, which converted its navy to fuel oil on its own technical merits Great Britain had to contend with the absence of domestic oil reserves and a corresponding dependence on foreign sources. Next best to domestic reserves was control over foreign reserves by a British company. As a result the British government gave diplomatic and even gunboat support to D'arcy in his Persian oil venture. With success came opportunity and with the prodding of the first Lord of the Admiralty, Winston Churchill, the government became a part-owner of the Anglo-Persian Oil Company.[46] Thus, a few days prior to the outbreak of the war the government found itself in the oil business.

Control over Persian oil did not mean neglect of the ancient Mesopotamian oil seepages that fueled the Eternal Fires of Nebuchadnezzar's furnace. There was, however, an important difference in the political miliéu. In Persia Britain had to contend only with Russia and in 1907, after 35 years of diplomatic ups and downs[47] both Powers came to an agreement on their

---

[46]"The government took up ₤2m. worth of shares in return for a voting interest of just over 50 per cent plus the power to appoint two directors with a veto on all strategic issues." Tugendhat, op. cit., p. 68.

[47]The first scene was enacted in 1872 with the Reuter concession which was cancelled the following year under pressure from St. Petersburg. Said Lord Curzon about the concession:
> "When published to the world, it was found to contain the most complete and extraordinary surrender of the entire industrial resources of a kingdom into

respective spheres of influence.⁴⁸ Mesopotamia, however, was part of the Ottoman Empire, and that led directly to the conflicting diagnoses of various self-appointed physicians on the state of health of the "Sick Man of Europe." In 1914, after a consultation of 26 years, the physicians agreed on a prescription and, more important, on their respective fees. The remedy consisted in the grafting of the Berlin-Baghdad railroad artery; the fee was a contribution to the newly formed hospital fund -- the Turkish Petroleum Company -- which was to be distributed as follows: 50% to Anglo-Persian, 23½% each to Shell and Deutsche Bank, and 5% to the invaluable head nurse Gulbenkian.⁴⁹ Soon after, the physicians had a falling-out. The patient made the mistake of taking sides and, even worse, of picking the losing

---

foreign hands that has possibly ever been dreamt of, much less accomplished in history."
George N. Curzon, Persia and the Persian Question (London: 1892), I, 480, as quoted in Benjamin Schwadran, The Middle East, Oil and the Great Powers (New York: Praeger, 1955), p. 14.

⁴⁸"In 1907 the Entente Cordiale was extended to include Russia, whose prestige and sense of security had been abased by here defeat in the Japanese War of 1904-5, and who was consequently more ready to compromise with her long-standing British rival. An Anglo-Russian Agreement was reached to 'obviate any cause of misunderstanding in Persian affairs' and to delimit the Russian and British spheres-of-interest in North and South Persia respectively, leaving a non-man's-land between them." George E. Kirk, A Short History of the Middle East from the Rise of Islam to Modern Times, Seventh Revised Edition (New York: Praeger, 1964), p. 94.

⁴⁹See Tugendhat, op. cit., Chapter 6.

side.⁵⁰ After ousting a member from their midst the physicians returned to their Hyppocratic duties and decided on the patient's amputation. The question was where, on what side of the gangrene?

This fable then brings us to the period under consideration. "The inter-war diplomatic and commercial jockeying for positions in the Middle East is a story that has been retold in numerous volumes from many points of view and of varying degrees of reliability."⁵¹ In view of the above we will refrain from a blow-by-blow description of this fascinating interlude and restrict the narrative to the minimum needed to highlight the variables of our analysis of the structures of the inter-war period.

The involvement of the Great Powers in the Middle East oil fields concerned the interrelated issues of territorial boundaries, spheres of influence, and oil distribution according to nationality and company. We will briefly discuss these issues as they evolved. The first point was the dismemberment of the Ottoman Empire. Any Soviet theoretical claims being disqualified by their weakness the issue was up to France and Britain.

---

⁵⁰No that the patient had much choice. "After all, the privileged positions of Britain and France in Lower Iraq and Syria respectively were encroachments on Turkish sovereignty; Russia, ever anxious to expand at the expense of Turkey, was constantly encouraging the Balkan, Armenian, and Kurdish nationalists; whereas Germany was the one Power whose interest it was to favour a strong Turkey." Kirk, op. cit., p. 97.

⁵¹Penrose, op. cit., p. 57.

> The Sykes-Picot Agreement of 1915 had arranged that the Fertile Crescent should be divided into four areas, two to be directly administered by France and Britain respectively, while the other two should be administered by Arab governments under the guidance and protection of France and Britain respectively. France's direct share was to be the Syrian coastlands and Cilicia, while her protectorate was to consist of Syria including the vilayet of Mosul.[52]

Now it just so happened that the vilayet of Mosul had the most promising oil field in the Ottoman Empire and that it had been included in the pre-war negotiations with the Porte in the concession to the Turkish Petroleum Company. As coincidence would dictate, when the British forces drove the Turks out of Mesopotamia in 1919, they also occupied the vilayet of Mosul. This then brought an interesting twist to Anglo-French relations.

> The [British] were aware that unless the oil could be brought by pipeline to a port on the Eastern Mediterranean, the Mesopotamian fields could not be exploited, and the road to the Mediterranean was in the hands of the French. The French were prevailed upon to give Mosul -- especially when it was militarily occupied by the British -- and to permit the oil to flow from the Mediterranean in return for a share of the oil.[53]

With the agrement in principle came an accord on details and the British transfered to the French the holdings of the Deutsche Bank. Thus satisfied the two mandate Powers prepared a draft peace

---

[52]Kirk, op. cit., p. 162.

[53]Schwadran, op. cit., p. 202.

treaty with Turkey which led to the 1920 San Remo conference.

The paragraph covering Mesopotamia read:

> The British Government undertake to grant to the French Government or its nominee twenty-five per cent of the net output of crude oil at current market rates which His Majesty's Government may secure from the Mesopotamian oil fields, the British Government will place at the disposal of the French Government a share of twenty-five percent in such company, the price to be paid for such participation to be no more than that paid by any other participant to said petroleum company. It is also understood that the said petroleum company shall be under permanent British control.[54]

Needless to say, the provisions of this agreement did not please other Powers. Germany's reservations on the subject could be easily discounted, but the Turkish protests had to be given more attention. Not that the Turks could do much about it, but there was that newly created organization -- the League of Nations -- with all its attachments to international ethics and legality. As a result the negotiations on the Iraqi-Turkish demarcations took time. But by far the largest outcry came from another source unconnected with the Turkish Peace Treaty or with the League, but with a hefty weight in international affairs: the United States. The result of the U.S. outcry was a classical repetition of the "open door" vs. the sphere of influence approach, adjusted to local circumstances.

---

[54]*Memorandum of Agreement between M. Philippe Berthelot, Directeur des Affaires Politiques et Commerciales au Ministaire des Affaires Etrangeres, and Professor Sir John Cadman, Director in Charge of his Majesty's Petroleum Department*, Cmd, 675, 1920, as quoted by Schwadran, *op. cit.*, p. 204.

> Beneath the diplomatic niceties both sides were driven by fear. The British, knowing that one of the most important props of their power and influence in the nineteenth century had been their indigenous coal and iron ore, were determined to get as much of the world's oil under their control as possible. . . . The empire had to have its own reserves, and the Middle East was the obvious place to search for them. To the Americans this looked like a sinister attempt to corner all the available supplies in order to hold them up for ransom. Despite their dominant share of the world's production, the Americans were convinced that their reserves were on the verge of running out.55

Which goes to prove that the unique characteristic of the oil industry, namely the uncertainties of exploration resulting in doubts about reserves, may have some ponderous foreign policy implications. So, with the prodding of the Secretary of Commerce, Herbert Hoover, a syndicate of the seven largest U.S. oil companies was formed in 1921 to represent American interests in the Middle East. Finally, after pressures from the State Department and from the major oil companies,56 " . . . the chairman of Anglo-Persian, asked the American syndicate in June 1922 to enter negotiations with a view to joining the Turkish Petroleum Company, and his invitation was promptly accepted."57

55Tugendhat, op. cit., p. 74.

56"Whatever the future might hold, America at that time was still the most powerful single force in world oil, with plenty of bargaining counters. Among the most important was Anglo-American, the former Standard subsidiary; it maintained close links with the New Jersey Standard and was second only to Shell in the British markets. It is not known whether the Jersey board ever actually threatened to cut off supplies, but they certainly considered the possibility, and news of the sort gets around." Ibid., p. 76.

57Ibid.

The negotiations themselves stand out as a watershed in the industry's history. For the first time they brought together all the world's largest companies, and the French at least were quite clear what this implied: 'It was the beginning of a long-term plan for the world control and distribution of oil in the Near East.' The results were the famous Red Line Agreement, which set the pattern for the Middle East concessions of today, and the Iraq Petroleum Company, the first of the great joint ventures which form the basis of the major companies' control over most of the world's reserves.[58]

Let us briefly discuss the issues, the actors, and their powers and interests. "There were three main points to be discussed: (1) the shareholding for each company; (2) the extent of the concession and the member companies' relationships with each other; and (3) the taxation arrangements."[59] After much ado the four interested parties came to the sensible solution of a division by four, less the 5% of Gulbenkian and a compensation for Anglo-Persian. The taxation arrangements dragged on mainly because of the conflicting interests between the four parties and Gulbenkian, but were finally agreed upon. It was the second point, the extent of the concession and the relationships between the companies that brought the major strife. In short, should the Turkish Petroleum Company be restricted to the promising Mosul vilayet or should it include Turkey and the remnants of its former empire, i.e., virtually the whole of the Arabian Peninsula. Closely related to the former

---

[58] Ibid., p. 82.

[59] Ibid., p. 83.

was the veto power of the individual partners over the actions of the others in the agreed upon concession area. Let us examine the respective positions, powers and interests.

The case of Gulbenkian was purely commercial. Having interests in only the Turkish Petroleum Company, he wanted the largest concession area and veto on his partner's operations. While obviously weaker than the others, he had a thorough knowledge of the industry, of the relevant governments, and of their relations; "his ultimate deterrent was to expose their deals and treaties in the courts, and it worked well."[60]

The position of the French was dictated by both commercial and political considerations. After the war Gulbenkian and Deterding had orginally proposed to bring the French into the Turkish Petroleum Company at the expense of the Deutsche Bank with the understanding that Shell would receive the French Government's share and act in its behalf. However, when Poincaré became Prime Minister he insisted on the formation of a French company. This was done in 1923, and the next year the newly formed company took the name of Compagnie Française des Pétroles, which today is "often referred to as an eighth 'international major'."[61]

The French wanted the concession of the Turkish Petroleum Company to cover the widest possible area in order to produce enough

---

[60] Ibid.

[61] Penrose, op. cit., p. 90.

oil to become independent of the international majors, and out of balance of payment considerations. But they also wanted a veto over their partner's operations in that area. A restriction of the concession to Mosul or even to Iraq would not only reduce their amount of possible oil production but, even worse, open up other possible fields in that area to their giant partners, thus leaving them in a weak position in the ensuing free-for-all. Finally, should the Iraqi fields prove sufficient to cover French needs, their veto over new production could be compounded to their control over the pipeline routes.

The British position was characterized by strength on both governmental and commercial levels. As the mandate or protector Power of most of the Arabian Peninsula Britain could control any future concessions in that area. By the same token the British companies, Anglo-Persian and Shell -- during the war Deterding had transferred Shell's headquarters to London -- could enter into any likely melee with the American giants on equal terms. Britain thus prefered to restrict the concession area to Iraq and to take its chances on eventual oil discoveries in the rest of the former Ottoman Empire. "But the French Government threatened to go to law, brought pressure to bear on London, and won the point."[62]

---

[62]Tugendhat, op. cit., p. 84.

The American position was characterized by a contradiction between government and oil companies at the verbal level and by coincidence of interests at the policy level. Government oratory was based on the "open door" policy and thus opposed to any restrictive practices that the US oil companies might agree upon. In practice both political and commercial considerations dictated the greatest possible freedom of action within the constraints imposed by Britain and France.

To make a long story short, Iraq granted a concession to the Turkish Petroleum Company in March 1925 for a period of 75 years in which it included the territory disputed by Turkey which was still under consideration by the League. Article 6 included a subleasing agreement of plots chosen by the Iraqi government, which it " . . . shall offer the same for competition, by sealed tender, between all responsible corporations, firms and individuals, without distinction of nationality, who desire leases."[63] As realized by all parties concerned, and as borne out by subsequent events, this article was a dead letter in view of the British control over the Iraqi government.[64]

---

[63]Schwadran, op. cit., p. 237.

[64]For an example of pro forma bidding, resulting in the granting of concessions to British interests, see the case of the Mosul Petroleum Company: Ibid., pp. 248-249.

103.

In December 1925 the League decided the Mosul boundary in Iraq's favor, and a month later a new treaty was signed between Britain and Iraq, thus outflanking Turkey on all fronts. As a result, in June 1926 a treaty was signed between Britain, Iraq and Turkey, thus transforming de facto occupation into de jure demarcation.[65] Drilling was started one year later and in October 1927 the first well was brought in with spectacular success, thus prompting the name change from Turkish to Iraq Petroleum Company and spurring renewed French cartographical efforts. Finally, in July 1928 the parties signed the Red Line Agreement, "the most famous example of an arrangement to curtail competition ever made in the international oil industry."[66]

> The Red Line encircled virtually the whole of the Arabian Peninsula. Within this area the companies agreed not to compete with each other for concessions nor to hold any individual concessions without first seeking the permission of their partners. Thus each company secured a veto over the activities of the rest.[67]

---

[65]"In addition to the boundary adjustments, the treaty provided that Iraq was to pay Turkey, for a period of twenty-five years, 10 per cent of the oil royalties from the area of Mosul. It also contained an option that Turkey might accept ₤500,000 in lieu of such royalties." Treaty between the United Kingdom and Iraq and Turkey regarding the Settlement of the Frontier between Turkey and Iraq, Cmd. 2679, 1926, as resumed by Schwadran, op. cit., p. 235.

[66]Tugendhat, op. cit., p. 84.

[67]Ibid.

So much for the narrative. Now let us analyze some of the mechanics of power. But first an evaluation of the random variable of geology. Let us assume for argument's sake that the known seepages were not in Mosul or Kirkuk but rather in Iraq's other prolific field: Rumaila, near the Persian Gulf. It takes little imagination to envisage the likely outcome. The British would have promptly signed a treaty with Iraq followed by a concession to Anglo-Persian which, leaving political matters to the British Government, would have developed the Rumaila fields at a speed comparable to its development of the Persian fields. To the eventual protests on the rights of the former Turkish Petroleum Company Britain would have evoked its higher responsibilities to its new mandate power and prepared cogent briefs studded with <u>dicta</u> and <u>obiter</u> <u>dicta</u> that justified its <u>fait</u> <u>accompli</u> with a categorical <u>rebus</u> <u>sic</u> <u>standibus</u>. In all likelihood the legal problem would hardly have arisen. And in the event of great pressure from the State Department Britain would have "opened the door" to sealed bids to be opened by its <u>protégé</u>

Nevertheless, to use the industry's motto, oil is where you find it -- which brings us back to the original situation. Let us inspect the ordnance of the contestants at the outset of the fray: Gulbenkian with his legal rights and "expertise," the French with their control over the access routes to the Mediterranean, the British with virtual control over the Arabian Peninsula, and the

105.

Americans with overwhelming political and commercial power, but with fear of exhaustion of their oil reserves. It has been seen that the outcome was a French victory, in the sense that they were able to parlay their control over the pipeline routes into roughly 25% of the future oil production of Turkey and most of the Arabian Peninsula and the veto power over operations in that area. But this tour de force was accomplished by means of flimsy transmission belts: continuing British control over the Arabian Peninsula and a persisting shortage of oil which would compel the giant American firms to abide by the French-inspired rules in order to gain access to the only known major Middle East oil field. It stands to reason that any change in the above would transform the French fortress into a house of cards. And change they did.

The first change occured in world reserves, starting with the United States. "In fact, 1921 turned out to have been the last year of shortage; from 1922 onwards the United States produced more oil than its refineries could handle, and the refineries in turn pushed out more products than the market could absorb."[68] Perhaps even more important were the new discoveries in Venezuela. "From 2m. barrels in 1922 . . . Venezuela's production rose to 9m. in 1924, to 37m. in 1926, and to 106m. in 1928, when it replaced Russia as the world's second largest producer."[69] The culmination came in

---

[68]Ibid., p. 80.

[69]Ibid., pp. 77-78.

1930 with the discovery of the East Texas fields, as a result of which ". . . the bottom fell completely out of the market as prices dropped from $1.30 a barrel to 5¢ a barrel,"[70] thus causing a trauma to the industry from which it has yet to recover.[71]

All these discoveries could not be without effect on the Middle East negotiations. "When they began in 1922 it was thought that nothing less than the regulation of the world's last great oil reserve might be at stake; when they ended six years later the industry's main problem had already become one of trying to curtail excess production.[72] As a result, when the Red Line Agreement was finally signed " . . . the membership of the American syndicate had dropped from seven to five. By 1930 there were only two left: Jersey Standard and Standard of New York, both of which had extensive markets in Europe and in the Far East."[73] But worse was yet to come.

We now turn to the second prop of the Red Line Agreement, British control over the Arabian Peninsula which, inter alia included various pledges not to "grant concessions within those territories to any Foreign Power, or to subjects of any Foreign Power, without the

---

[70]Ibid., p. 93.

[71]"Nothing, said Joseph Schumpeter, is so durable as a folk memory. And the influence on oil industry thinking of the catchwords 'Remember East Texas in '30; we'll have 10 cents a barrel if we don't watch out!' is extraordinary." Adelman, The World Oil Outlook, op. cit., p. 72.

[72]Tugendhat, op. cit., p. 82.

[73]Ibid., pp. 86-87.

consent of the British Government."[74] On the surface, the combination of British political control over the Arabian Peninsula resulting in the exclusion of non-British companies in that area, and the inclusion of the two major British oil companies in the Red Line Agreement seemed to make the latter watertight. Nevertheless, in 1933, with the oil hardly dry from the signature, the Red Line Agreement had become worthless outside of Iraq.

The first crack in the structure came in what today is Saudi Arabia which after the partition of the Ottoman Empire was fragmented among various feuding parties. By 1921 the situation had crystalized to the main rivalry between Ibn Saud -- the Sultan of Najd -- and King Husain of the Hijaz. While Britain subsidized both rivals, the subsidy to the former was conditioned to his holding the peace with the latter. "But the old King, with greater consistency than worldly wisdom, broke with Great Britain, mainly over the political disability imposed on the Arabs of Palestine by the Balfour declaration and the Mandate. Refusing to compromise on the point, he forfeited Britain's support and subsidy."[75] To make matters even worse, King Husain, who in 1916 had proclaimed himself King of the Arabs , proceeded to anoint himself as Caliph, thus causing

---

[74] Formulation from Article IV of the 1915 Treaty between Great Britain and Ibn Saud, at that time the Sultan of Najd, in Schwadran, op. cit., p. 285.

[75] Kirk, op. cit., pp. 161-162.

". . . serious disturbances in Egypt and India as well as in Arabia."76 At that point Britain stopped paying the monthly ₤5,000 to Ibn Saud, thus giving him a free hand to dispose of his rival. After a victorious invasion Ibn Saud became King of the Hijaz, Najd, and its dependencies in 1926, and was recognized as such by Britain in the 1927 Treaty of Jidda. As a result of this treaty the concession restrictions were voided for the later renamed Kingdom of Saudi Arabia.

For reasons of chronology we now turn to the activities of Major Frank Holmes, known as "Abu el Naft" -- the father of oil -- whose faith in his oil finding abilities, British nationality and shortage of capital produced an explosive combination that triggered the avalanche that burried the Red Line Agreement. As opposed to the then prevalent geological opinion Holmes was certain of the existence of large oil reserves in the Persian Gulf. His congeniality and British nationality enabled him to clinch concessions from Ibn Saud in the Hasa region, and from the Sheiks of the island of Bahrein and Kuwait in the early 1920s, but his lack of capital to develop these areas of the Persian Gulf forced him to search for sponsors. First he tried the three major British companies. "When all three returned negative replies Holmes reluctantly turned to the United States."77

---

[76] Schwadran, op. cit., p. 287.
[77] Tugendhat, op. cit., p. 90.

After much pondering Gulf decided to take the risk and in 1927 it took over Holmes' concessions, only to be vetoed by the IPC directors. The Hasa concession having lapsed for lack of payments Gulf persuaded an outsider, Standard of California, to take over the Bahrein concession. This immediately set off a new round of "nationality clause" vs "open door" discussions between the Foreign Office and the State Department. Finally, the Foreign Office agreed in principle to the inclusion of American interests and suggested that since Holmes' syndicate " . . . the Eastern and General Syndicate was still the concessionaire, the syndicate should negotiate with the Colonial Office."[78]

> Negotiations between the Colonial Office and the Syndicate were carried on for about a year and a half. On June 12, 1930 the Syndicate signed an agreement with the Sheikh, and transferred it to the Bahrein Petroleum Company, a Canadian subsidiary of the Standard Oil Company of California.[79]

While the conditions under which Standard of California was allowed to operate were rather harsh, it still represented a successful penetration by American interests in that area.[80] Drilling was soon started and oil was found in 1932.

---

[78] Foreign Relations 1929, III, 80-81, in Schwadran, op. cit., p. 373.

[79] Ibid., p. 374.

[80] "Here again the prompt and positive action of the State Department has secured results favorable to an American-owned company. By securing the entry of American oil interests into Bahrein, the way was paved for some American interests to obtain concessions in nearby Arabia." Ibid., citing U.S. Senate, American Petroleum Interests in Foreign Countries, pp. 23-24.

The major companies in Iraq heard the news with
horror. They thought they had secured a monopoly
of the Middle East's resources, and now here was a
company outside of their charmed circle finding a
new and independent source of supply. Moreover, if
there was oil in Bahrein there was a good chance
that their expert's reports on the rest of the
Persian Gulf could turn out to be equally wrong,
and a rush for concessions was bound to begin.[81]

And here is where Britain's loss of political control over what now was Saudi Arabia was acutely felt. For when IPC went to bargain with Ibn Saud the latter had freedom of choice. The King asked for some gold in advance, IPC demurred " . . . but California Standard realized that this was not an occasion for haggling. Within forty-eight hours it had deposited the full quantity of gold, and clinched the contract."[82]

The rush for concessions spread further to Kuwait. The possibility that there might be oil in Kuwait was recognized early by the British government. In 1913 the Sheikh agreed that "we shall never give a concession in this matter to anyone except a person appointed by the British Government,"[83] and the British had excluded Kuwait from the area of the Red Line Agreement. Nevertheless, the British failed to take a concession for themselves, leaving the field to

---

[81]Tugendhat, op. cit., p. 91.

[82]Ibid., p. 92.

[83]C.U. Aitchison (ed.), A Collection of Treaties, Engagements and Sanads Relating to India and Neighboring Countries, XI, 2640265 (Delhi, 1933), in Schwadran, op. cit., p. 384.

Holmes and by indirection to Gulf who,"After being prevented from being allowed to go into Bahrein it had resigned from the Iraq Petroleum fields, and was thus cut off from both the new Middle East fields."[84] The negotiations between Anglo-Persian and Gulf cum Holmes lasted for two years with large assist from their respective governments. "By the end of 1933 both sides were ready to compromise, and it was agreed to set up a jointly owned Kuwait Oil Company registered in London."[85]

Let us briefly sketch the events leading to the dissolution of the Red Line Agreement. In 1934 the first tanker called in Bahrein. Next year Standard of California sold half of its Bahrein interests to Texaco, and the year after that, half of its Saudi Arabian interests as well. In 1938 commercial fields were discovered in Saudi Arabia and Kuwait. Some production was started in Saudi Arabia in 1939, but the rapid development of both fields had to wait until 1946. The two remaining American companies in the IPC, Standard of New Jersey and Socony Vacuum (formerly Standard of New York, presently Mobil), were chafing under the restrictions imposed upon them by their partners and wanted to gain access to the newly developed Middle East fields.

---

[84]Tugendhat, op. cit., p. 91.

[85]Ibid., p. 92.

113.

The war gave them the opportunity by disrupting all the Iraq Petroleum arrangements. With the fall of France the Compagnie Française and Gulbenkian, who remained in Vichy, were declared enemy aliens, and cut off from their oil, while even the allied groups could not carry on normally. Jersey Standard and Socony Vacuum claimed that as a result the Red Line Agreement had been dissolved, and that to reinstate it would be contrary to the American anti-trust laws.[86]

This brought on a series of swaps and arrangements, whereby these two companies bought into Aramco of Saudi Arabia (present ownership: 30% each Standard of California, Standard of New Jersey and Texaco, and 10% Mobil) and signed long-term contracts with Anglo-Iranian (formerly Anglo-Persian, persently British Petroleum) for the purchase of Kuwait oil, while Shell did the same with Gulf. "Thus by the end of 1947 the interests of all the major Anglo-American groups in the Middle East had been brought into a happy state of harmonization."[87]

Needless to say, this was not pleasing to the French. But this time their elaborate structure worked against them. To put it differently, the flow of their transmission belts had been reversed. The production from the giant fields of Saudi Arabia and Kuwait was outside of French control and reduced the Anglo-American dependence on Iraqi oil. The Free French had in 1941 granted the

---

[86] Ibid., p. 95.
[87] Ibid., p. 97.

independence to Syria and Lebanon. In 1946 the French withdrew their troops under intense British, American and United Nations pressure, thereby losing the control over the Iraq pipelines to the Mediterranean. As a result the French position in Middle East oil had been reduced to legal rights in the IPC which were offset by the Anglo-American threat to delay production from the Iraq oil fields. In desperate need to rebuild its shattered economy France agreed to the dissolution of the Red Line Agrement in return for expanded production from Iraq. In regards to Gulbenkian, he threatened to go to court with this "expertise," scared his partners, and was promptly paid off.

We have so far discussed at some length the actions and interactions between oil companies and their respective home governments for the control of Middle East oil reserves. The role of the Middle East governments was purposely omitted in view of their limited freedom of action. In Venezuela the dominant actors were the oil companies and the local government. This represents a constellation of interests and a set of power mechanics found in our present structure, and will be discussed in this context. Suffice it at this point to note that in 1929, after the dust had settled, more than 98 percent of Venezuelan oil was produced by Shell, Gulf and Standard of Indiana, and that in 1939, after a series of swaps and purchases, 92 percent was produced by Standard of New Jersey and Shell.[88]

---

[88]Lieuwen, op. cit., pp. 44, 85.

Having noted a similar concentration as the one that emerged in the Middle East we can now turn to the other significant aspect of the inter-war international oil industry structure -- the formation of the cartel. For this we must revert to the summer of 1928, to the time of intense bargaining between the oil companies that led to the Red Line Agreement.

> The major companies' efforts to establish a monopoly -- or rather an oligopoly -- over the production of oil were not surprisingly accompanied by a similar attempt to control its distribution. The aim was to establish an international cartel to regulate prices and competition in such a way as to guarantee a profit to all the members.[89]

It has already been seen that oil cartels had flourished in pre-World War I Europe. But these were fluid in nature, rarely crossed national boundaries, and were usually both the result and the prelude to price wars. These arrangements lasted with some variations into the 1920s at which time the structure of the industry had changed to a degree that could make price competition a ferocious affair. In the late twenties' the vertical and horizontal integration of the giants had been largely accomplished. To the backward integration into crude was added the forward integration of the modern sales network, up to the drive-in filling stations. In view of the different advantages and market shares of the giant companies in various countries the scene was set for an international price

---

[89]Tugendhat, op. cit., p. 97.

war. The spark that lit the fuse was Shell's resentment of Standard of New York's sales to India of Russian oil. As the Soviet Government had failed to compensate for the seizure of Shell's production facilities, Deterding vented his spleen over what he considered the sale of stolen oil by starting a price war in India.

> What began as a purely local struggle quickly spread to Europe and the United States. More and more countries were dragged in, and the other companies began to suffer through being forced to cut prices in order to retain their positions. The fight could not continue indefinitely, and eventually at the end of 1928 Shell, Standard of New York, and Anglo-Persian reached an agreement on how the Indian market should be shared out. By itself this would not have been particularly important, but it became the first step in a much wider settlement. All the international companies had been thoroughly frightened by the way an apparently local dispute had spread across the world and felt that it could not be allowed to happen again.[90]

As a result, after due preparations and deliberations, the heads of Shell, Standard of New Jersey, and Anglo-Persian had signed the so-called "As Is" or Achnacarry agreement -- the name of the Scottish castle where the conclave was held, " . . . from which the agreement -- the the delight of several generations of journalists -- derived its sonorous if puzzling name."[91]

The "As Is" Agreement involved more than market share and prices. This alone would well have merited Frankel's description

---

[90]Ibid., pp. 98-99.

[91]Frankel, Mattei: Oil and Power Politics, pp. 84-85.

of a cartel as a "middle-class version of the trust."[92] This cartel tried to shelter itself from Schumpeter's "perenial gale of creative destruction" by eliminating the " . . . competition which commands a decisive cost or quality advantage and which strikes not at the margins of the profits and the outputs of the existing firms but at their foundations and their very lives."[93] This was accomplished by a pooling of facilities and by setting an international price for oil.

> They were to combine their interests and share each other's facilities -- refineries, storage, tankers, and the rest -- in order to present a united front against companies trying to break into new markets, price cutters, and other disturbing elements. ... In essence therefore the companies were agreeing to function virtually as joint ventures, with each shareholder contributing a certain proportion of oil and capital investment and receiving a share of the profits in return.[94]

In regards to market shares " . . . with true Anglo-Saxon pragmatism it was agreed that, since one could not find a logical formula on which the relationship of the several participants could have been based, all concerned should recognize each other's relative standing on the market and say to each other 'Let us keep the position as (it) is'."[95] This was well in keeping with the

---

[92]Ibid., p. 84.

[93]Joseph A. Schumpeter, *Capitalism, Socialism and Democracy* Third Edition, Harper Torchbooks (New York: Harper & Row, 1962), p. 84.

[94]Tugendhat, op. cit., p. 84.

[95]Frankel, op. cit., p. 84.

equally sensible solution of distribution of shares in the Red Line Agreement.

> Naturally all this involved working out a system of quotas for each market and agreeing on a formula for fixing prices. The distribution of quotas, the allocation of transport, the fixing of freight rates, and various other administrative matters were made the responsibility of a central association in which each member was represented.[96]

We can neglect the market sharing and price fixing in the various countries because of their purely local character and turn our attention to the international pricing of oil.

> The problem of prices was solved by the formulation of the Gulf-plus pricing system. This laid down that the price of oil should be the same in every export center throughout the world as in the American ports along the Gulf of Mexico. But the final cost at the point of delivery should vary depending on its distance from the Gulf of Mexico and on whether or not the buyer was a member of the cartel.
>
> Ordinary commercial customers and oil companies outside the cartel had to pay the basic Gulf price plus the cost of shipping the oil from the United States to the point of delivery, wherever that happened to be. . . . When companies within the cartel were selling to each other profiteering on this scale could not be allowed. The Gulf of Mexico price was still used as the peg for pricing all over the world, but any freight savings achieved by drawing supplies from another source were shared equally between buyer and seller.[97]

---

[96] Tugendhat, op. cit., p. 102.

[97] Ibid., pp. 102-103.

There are two elements of interest in this pricing system: its rationale and its implications. "Basically, it was a logical method for pricing oil in world markets under the supply conditions and industrial structure prevailing prior to the Second World War. The oil supply pattern was characterized by three principal features:"[98]

"a) The United States oil industry was by far the largest in the world.

b) U.S. oil exports from the Gulf Coast covered a large part of the world's demand, with other Western Hemisphere and Middle East supplies still in the process of rapid expansion and development.

c) The U.S. Gulf was practically the only place where importers could obtain supplies and spot cargoes on the open market to cover any likely requirements."[99]

Fair enough, but this does not explain why oil drawn from the Persian Gulf and delivered to Karachi had to pay a "phantom" freight from the Gulf of Mexico. The explanation to this is that the Middle East (and Venezuela) oil fields were largely controlled by the majors whose main investments were in the United States. Therefore it became no only possible but also profitable to maximize the exports from their largest facilities " . . . and if that meant

---

[98]Frank, op. cit., p. 10.

[99]Ibid., pp. 10-11, quoting Walter J. Levy, "The Past, Present and Likely Future Price Structure for the International Oil Trade," Third World Petroleum Congress, Proceedings, Sec. X. (Leiden: E.J. Brill, 1951), p. 116.

charging a high price for Middle East and Venezuelan supplies, then so much better for profit margins."[100] As Tugendhat continues:

> If the new fields in the Middle East and Venezuela had been discovered by companies without any assets in the United States the situation would have been quite different. There would have been a repetition of what happened in the 1880s when Russian oil flooded into Europe to break Standard's domination, and forced the company to reduce the price of its American exports.[101]

There were many advantages for the cartel members which later included most of the major companies and which, outside of the Soviet Union and the United States, was international in scope. It was in this spirit that the previously mentioned Middle East and Venezuelan swaps, purchases and joint ventures took place that allowed for a rational exploration of these large new discoveries. Perhaps more important, these arrangements permitted the orderly retrenchment and curtailment of supply during the world depression.

As cartels and commodity control schemes go it was rather modest when compared with the British rubber restriction scheme covering the output of Malaya and Ceylon, or the Cuban sugar and Brazilian coffee arrangements, and definitely humble when compared with the monopolistic practices of the international copper control schemes which prompted a buyer's revolt.[102] One should hold in perspective that " . . . the evolution of cartels in manufacturing

---

[100]Tugendhat, op. cit., p. 102.

[101]Ibid.

[102]Rowe, op. cit., pp. 121-26.

industries, and in some mining industries, took place earlier than it did in agriculture and other mining industries. Put very summarily, the former date a long way back before World War I, while the latter is a post-World War I development."[103] But more about this later. In the meantime some figures to fix ideas for the discussion of the present structures:

TABLE II

The Main Oil-Producing Countries in 1938

Country	Millions of Barrels
United States	1165.6
Venezuela	199.4
Iran	73.4
Mexico	39.8
Iraq	31.8
Total	1510.0

Source: Tugendhat, op. cit., p. 112. Soviet Union and Rumania excluded, both countries accounting for about 13 percent of world production.

---

[103] Ibid., p. 121.

IV. THE PRESENT STRUCTURE

The present structure of the international oil industry is one of relative fluidity. Its main actors are the international majors and the oil producing countries. Its hallmark is the dependence of the industrialized countries outside of the United States and the Soviet Union on Middle East oil, and the resulting dialectics between economics and politics on the national and international stage.

The importance of oil during World War II needs hardly any elaboration. In this motorized world conflict oil was one of the main sinews of war, and thus loomed large in the strategies of both Allied and Axis Powers. For the latter it meant increased reliance on synthetic oil and great strategic thrusts for the control of oil reserves, witness the campaigns of Germany in North Africa and Southern Russia, and of Japan in present Indonesia and Burma. For the former it meant resistance to these thrusts. By the same token oil presented an ideal bottleneck for the contending Powers. As a result, tankers, refineries, and, on occasion, production facilities were favorite targets of submarines and bombers. While it is easy to overemphasize the role of oil by simple reductionism, there is no doubt that the outcome of the "battle of oil" was an important factor in the outcome of the war.[104]

---

[104]For an exhaustive as well as exhausting account of the strategies, tactics and technologies encompassing the role of oil in World War II, see Maurice Lévêque, Le Pétrole et la Guerre (Paris: Debresse, 1952).

> In many ways the Second World War marks a
> turning point in the history of the industry,
> for after the war there was a steady decline in
> the control exercised by the major companies, which
> was not apparent in the ten years or so following
> its end but which became unmistakably evident in
> the later 1950's.[105]

The most important change was the loss of military and political control by the European Powers over their respective colonies and mandates, and the corresponding ushering of new actors on the stage, each with his own economic and political interests.

> The international petroleum industry could
> hardly be immune to these far-reaching psychological,
> political, and economic changes (sometimes referred
> to as 'de-colonialization') which marked the post-war
> decade and were, in part at least, the result of the
> impact of war on le tiers monde.[106]

The major effect of the de-colonization of the Afro-Asian countries on the oil industry was a change in the rules of the game. Hence, the economic and political interests of the oil producing countries had to be taken into account. Let us now examine how these interests were fitted in the evolving structure of the industry. Throughout this brief discussion emphasis will be stressed mainly on the economic aspects; the national and international political aspects have been briefly touched in the previous chapter and are the main consideration of Part II.

---

[105] Penrose, op. cit., p. 62.

[106] Ibid., p. 63.

The discussion of the inter-war structure led us to the formation of a cartel of the major oil companies that lasted well into the 'forties. "Anyhow, a world war and a new and more ambitious U.S. Anti-Trust doctrine, which now affected all American corporations also in their dealings abroad, disposed of As Is and all its works."[107] But after nearly two decades of mutual "assistance" the main features had been firmly implanted. These consisted in the control over the major world low-cost oil fields and close cooperation in transportation, refining and storage. "All this did not amount to a classical cartel, the set-up made it unnecessary to have a cartel; the main points of potential conflict having been eliminated, the Seven could rely on the inevitable parallelism of interests of firms in an oligopoly."[108]

In 1943 the United States became a net importer of oil and the former rationale for the Gulf-plus pricing system ceased to exist. Nevertheless, there was a need to balance inter-regional crude oil prices. Not that crude oil prices mattered greatly in inter-affiliate transfers but there still was a marginal free market which required inter-regional equilibrium. Besides, posted prices for crude oil meant dollar payments for importing countries, and were thus of great interest to the Marshal Plan administration.

---

[107]Frankel, op. cit., p. 85.

[108]Ibid., p. 87.

After constant prodding from the E.C.A. the majors agreed to post Middle East prices that " . . . were determined by the value such crude had delivered to New York less average cost of tanker transportation from the Persian Gulf."[109]

> The international majors could not determine their prices in the Middle East without considering the impact on Venezuela and, before the imposition of mandatory import controls, the United States. The eruption of inter-regional price wars was thus just as unlikely as a price war within the Middle East.[110]

The net effect of this maneuver was the tying of low-cost Middle East oil to the high prices of U.S. oil, thus giving the majors a magnitude of profit that allowed them to invest into large-scale "down-stream" facilities from retained earnings. Unfortunately for the majors, the difference between low-cost and high price did not go unnoticed by the oil producing countries. This started off a round of discussions which culminated in the so-called "fifty-fifty" split.

> Originally, the rulers and governments of these countries got some modest initial payments on granting a concession, usually combined with some rental for the territory covered by the concession; thereafter, if oil was found in quantities which justified exploitation, amounts varying with the quantity of oil, i.e., a so-called royalty, became payable too.[111]

---

[109] Ibid., p. 88.

[110] Frankel, op. cit., p. 131.

[111] Ibid., p. 91.

The first change in this system occured in Venezuela in 1943. The Venezuelan government had long pressed for a greater share but its demands were not heeded due to its bargaining weakness. "The problem remained unresolved until the middle of the war, when the allies' desperate need for oil gave the Government . . . a much stronger bargaining position."[112] The result of this change was the first of the so-called "50-50" agreements whereby the companies and the government shared the profits arising from the difference between cost and price.

In the Middle East some of the concessions had included some form of profit-sharing, but this was " . . . swiftly abandoned when margins virtually disappeared during the economic depression of the 'thirties, in favor of the much safer system of straight royalty payment."[113] However this might be, the past policy of not sharing losses did not hamper the desire of sharing present gains. Thus the system was inaugurated whereby production costs and royalties were subtracted from the posted export price and the remaining profit split in two. "In 1950 Saudi Arabia became the first to implement the new system, and by 1952 all the other important produces in the area had followed suit with the exception of Iran."[114] This system

---

[112]Tugendhat, op. cit., p. 131.

[113]Frankel, op. cit., p. 92.

[114]Tugendhat, op. cit., p. 134.

did not unduly depress the major companies as they were able to offset the new payments to the producing countries against their tax liabilities at home. As Frankel remarks: "One cannot help noticing that international operations, with all the dangers which navigation in treacherous currents involve, do at times provide ample compensation."[115]

The change in taxation procedures had the effect of increasing the producer governments' take and of pegging their income to the level of posted prices. While this technicality was to bring future complications nobody seemed to grudge the new distribution of the increasing pie. But trouble was rumbling in the distant horizon. This was due to two main economic reasons: overabundance of supply in relation to demand, and the arrival of new competitors.

The increased production of low-cost Middle East oil had had little effect on the world price of crude oil due to a set of extraneous political circumstances which either increased demand or curtailed supply. But after Korea, the Iranian nationalization episode (which is discussed in Chapter IV), and the Suez crisis, ". . . the day of reckoning . . . was finally at hand . . . The world industry has been operating since that time in a strong buyer's rather than a seller's market. Moreover the strength of this buyer's

---

[115]Frankel, op. cit., p. 93.

market has grown continously under the impact of worsening over-

supply conditions."116

According to Tugendhat, "One of the strengths of the
free-market system is that it is self-correcting, and this is as
true of the international oil industry as any other. Whenever there
is a seller's market and prices are high new companies are attracted
to the business, leading to the discovery of new fields, more competition, and lower prices."117 If by free market system we understand
lack of control, the above statement adequately reflects the conditions
that led to the arrival of the newcomers: the companies of the
consuming government and the American independents.

> The organizations from the consuming countries
> were rather slower off the mark than the Americans
> for the obvious reason that after the war their
> Governments were desperately short of capital.
> The Germans and the Japanese were, of course, in
> a hopeless position to do anything, and at the first
> the running was taken up by the French and the
> Italians.118

With the somewhat painful experience in Iraq behind them,
and with only a small stake in the newly formed Iran consortium the
French decided that it was safer to look for oil in their own
empire. Accordingly, the wholly-government-owned B.R.P. searched

---

[116]Frank, op. cit., pp. 92, 61.

[117]Tugendhat, op. cit., p. 146.

[118]Ibid., p. 150.

and found oil and gas in the Sahara. Despite the Algerian revolution, " . . . the Saharan oil and gas reserves were developed extraordinarily quickly, and when Algeria achieved its independence in 1962 the new Government took over a thriving industry with production running at 20.7m tons a year."[119]

The Duce's *mare nostro* having been reduced to the Italian beaches, Mattei of state-owned E.N.I. had no other alternative than to compete with the majors. Lacking the financial strength to compete on equal terms he made a virtue out of necessity by offering "true partnerships" to the producing countries. This took the form of a joint company with the National Iranian Oil Company with the latter receiving 75% of the proceeds. This venture found no oil but set the precedent for new types of concessions.

The interests of the American independents (i.e., firms of various size outside of the rank of the majors) in foreign ventures began with the import of cheap foreign oil into the U.S. that threatened their respective markets. First in Canada, then in Venezuela, and then gradually in the Middle East -- in Iran via the Consortium, in the Trucial Shekdoms, in the offshore areas of the major producers, and finally in Libya -- they began producing mainly for their U.S. markets. In 1957 the U.S. imposed "voluntary"

---

[119]Ibid., p. 151.

controls and in 1959 it established the present quota system which de facto excluded most of the independents. Having lost their markets for their new found oil the independents turned to Europe. "As a result, prices in Europe dropped sharply, and in Germany, for instance, heavy fuel oil fell from a peak of DM 142 a ton in February 1957, to DM 60 in the second half of 1959."[120]

"When prices began to collapse under the weight of oil coming on the market it was assumed in some quarters that the major companies would resume their efforts to form an international cartel."[121] But, as Frankel wrote, "It's a long way to Achnacarry."[122] "Instead of selling their oil at the official posted prices, they started offering unofficial discounts and cutting the prices of their refined products."[123] Loss of profits required readjustments, but as the companies found out, their freedom of action had been curtailed by conflicting government interests.

> The precariously balanced equilibrium of the international industry depended on successful strategy in three critical areas: the relations with the producing countries; the attitude of importing countries towards supply and price; and the actual amount of crude oil coming on the market in relation to demand.[124]

---

[120] Ibid., p. 158.

[121] Ibid., p. 156.

[122] P.H. Frankel, Oil: The Facts of Life (London: Weidenfeld & Nicholson, 1962), p. 18.

[123] Tugendhat, op. cit., p. 157.

[124] Penrose, op. cit., p. 69.

In the absence of effective control over supply, and with supply vastly exceeding demand, competitive forces brought down the price of crude and of refined products. But the posted price for crude was tied to the income of the producing countries. Consequently the companies maintained the crude posted prices, but compensated their losses by juggling their books to show downstream losses in the consuming countries. The consuming countries resented not only the ensuing loss of fiscal revenue but also the fact that it was not compensated by a reduction of the foreign-exchange cost of their oil imports. Acceding to pressure the companies reduced the posted prices for crude. While the reductions were not too large their impact was a shock to the producing countries. Their reaction in 1960 was the formation of a producer's cartel, the Organization of the Petroleum Exporting Countries.

The first thing that the newly formed OPEC demanded was the restoration of the posted price levels. This it has failed to accomplish. It did, however, transform the posted prices into "tax reference" prices that, for all intent and purposes, could not be unilaterally reduced by the companies, thus insulating the producing countries from the vicissitudes of the free market. After years of collective bargaining between OPEC and the majors, agreement was reached on royalty expensing -- a formula that gives a larger share of the profits to the producing countries. As *quid pro quo* OPEC agreed to "marketing allowances" (discounts) which were to be progressively eliminated according to a fixed schedule.

The other net benefit that accrued to the producing countries has been a radical change in the concession patterns. These have been achieved by unilateral bargaining with the companies, by unilateral action -- read legislation -- and by the use of the "most favored country" clause in the new concessions. Perhaps the most important aspect has been the large-scale relinquishing and/or confiscation of acreage held by the majors, thus giving the producing countries a possible source of additional revenue and enhancing their bargaining power.[125] The role of OPEC in these proceedings is difficult to judge. While there are some cases where collective action has brought pressure to bear on recalcitrant companies the record is by no means clear. Suffice to say, that the mere fact of OPEC's existence was an effective bargaining point in the various bilateral negotiations and a point of strength in the instances of unilateral actions.

According to Hartshorn, "It could be argued that all OPEC had achieved had been on behalf of its Middle East members, not the others. Alternatively, you could rationalize it a little sourly by

---

[125]For many newcomers companies, ". . . the expected costs of finding and developing new reserves of crude oil were below the expected cost of buying their requirements from the market. . . . Thus a 'sellers' market' in oil concessions existed side by side with a 'buyers' market' in crude oil and products." Ibid., p. 77.

saying that anything which increased the tax-paid cost of Middle East oil helped Venezuela by making its oil to some extent more competitive."[126]

Leaving OPEC aside, and partly as a result of the large-scale relinquishment of acreage by the majors, the last decade has ushered a proliferation of new types of agreements. With the exception of Soviet dealings with third parties the new agreements are mainly between the various national oil companies of the producing countries[127] and either the established and/or newcomer companies, or the consuming governments. The newcomer companies include some American independents and various consuming country companies of a mixed degree of private and government ownership. The forms that these new agreements have taken include among others the so-called 75-25% concept, equity participation, and a mixed bag of "partnerships" with assorted degrees of capital, risk taking and vertical integration. The intergovernment agreements have taken place either on a normal cash basis with various degrees of discount, or on a bilateral barter basis, usually with Communist countries.

All the above new types of agreements may perhaps be an omen of things to come, but their importance on the present structure

---

[126] Hartshorn, op. cit., p. 343.

[127] "These companies are the children, so to speak, of public opinion, and are viewed as symbols of the national aspiration to gain a place in the sun on the international scene." Ashraf Lufti, OPEC OIL, Middle East Oil Monographs No. 6 (Beirut: Middle East Research and Publishing Center, 1968), p. 37.

135.

of the international oil industry has been negligible. The main actors continue to be the international majors and the producing countries, and it is their present and likely future relations that we now turn. Let us first examine world supply. "The great paradox of the international oil industry is that ever since the mid-1950s it has been facing a world surplus and declining prices, yet exploration is still its most important single activity."[128] The reasons for this paradox are commercial and political. In regards to the former:

> The industry is not a monolithic whole; it is a collection of different companies with conflicting interests which have to compete with each other in order to earn their profits. Some . . . have more oil than they need. But this is no consolation for Shell . . . nor to . . . such as the Italian ENI and French ERAP. For these groups the fact that there is a world surplus of oil is quite immaterial. It belongs to other people, and they want their own. Not even the reserve-rich companies can afford to rest on their laurels.: as . . . the key to profitability in the international oil industry is to have a source of supply as close as possible to every market.[129]

The political reasons for this paradox are based on the past experiences with the Middle East, including North Africa, which, outside of the United States and the Soviet Union, is blessed with 75 percent of the world reserves and 46 percent of the production.[130]

---

[128]Tugendhat, op. cit., p. 165.

[129]Ibid., pp. 166-68.

[130]Ibid., p. 165.

Having experienced three disruptions of supply since World War II, the majors are most likely to persist in their exploration of new areas.

The success of new explorations will depend not only on the geological roulette but also on the on- and offshore technological developments described in the previous chapter. At the time of writing the most promising area seems to be a breakthrough in the adaptation of oil technology to Arctic conditions. A successful overcoming of the vast problems of transportation: pipelines, icebreaker tankers (the historical trip of the Manhattan through the North West Passage is only a beginning) and -- even more taxing -- ocean terminals, will not only allow full exploitation of the Alaska reserves but also provide the additional incentive to explore the rest of the North American land mass beyond the Arctic Circle.[131] This area appears to be geologically promising; it also has a transportation advantage to the major consuming centers over the Persian Gulf.[132]

From the discussion so far we can thus safely assume continuing exploration in new as well as traditional areas. Considering

---

[131]This, of course, is also true for Siberia. Its location, however, would restrict its export impact to Japan and -- perhaps -- China.

[132]One can hardly help to speculate on the eventuality of Canada becoming a major oil exporter. The foreign policy of oil vs. the domestic and foreign policy of wheat?!?

the vast world oil resource base and further technological developments one may assume a degree of success in exploration roughly proportional to the capital invested. Speculations on the characteristics of the to be discovered oil fields must remain speculations. My personal guess, based mostly on history and partly on recent discoveries (Alaska, Indonesia, East China Sea) is that while there may not be future Middle Easts there will be enough large-scale discoveries of low-cost oil to maintain the present range on Middle East posted prices. If these assumptions are substantially correct they all add up to one thing: vast proven reserves of low-cost crude capable of supplying oil in quantities far above demand.

It is then within the above framework that the future relations between the majors and the producing countries will most likely evolve. Let us examine the respective bargaining positions. The main weapon of the producing countries is sovereignty. Without entering into a discussion of politically meaningless legal arguments and rationalizations this means that they have the power to act as they please within the confines of their borders. In its extreme forms it includes the rights to tax the companies out of existence at a moment's notice, to nationalize and to confiscate, and even to hold hostages. By the same token, and assuming no intervention by their home governments in their affairs, the companies have the right to restrict or even to cease export from any given country and, if acting in concern, also the power to prevent the sale of all but marginal quantities of oil on the major markets. Within these extreme positions the actual bargaining takes place.

Let us first examine the question of nationalization. According to Lufti, whose ardent advocacy of the Arab and OPEC cause is -- in this respect -- tempered by a large sense of realism:

> The fact that an exporting country has huge reserves under the surface of its territory can only mean just that, no more no less. It is one thing to own these reserves and quite another thing to have the ability to extract them profitably under present international market conditions. With the current abundance of potential crude supplies, what counts in the first instance is not the ownership of the source -- there are quite a lot of sources now -- but the ownership of the market outlets; the gasoline stations and the long-term contracts for supply of refined products. In the second rank of importance are the processing plants and facilities, ownership of which is a guarantee for the flow of the crude oils needed by those plants and facilities.[133]

As a result of the above any country attempting nationalization would find that " . . . there would be bound to be a sharp drop from which that country could scarcely be expected to recover if the present marketing conditions continued to prevail and if that country's act of nationalization were kept isolated."[134] In other words, " . . . any individual initiative can be little short of suicidal."[135] Collective nationalization might have a

---

[133]Lufti, op. cit., p. 39.

[134]Ibid., p. 42.

[135]Ibid., p. 44.

chance, but this would require the prior act of joint production programming, to which we now turn our attention.

The problem with excess supply and competition is that these lead to discounts on the level of posted prices. "If these discounts continue to grow, the oil companies will sooner or later find themselves in the position of either having to reduce tax prices or of being forced out of business. OPEC is opposed to either alternative . . ."[136] As a result there is " . . . one and only course left open to OPEC member countries -- namely a joint production program requiring the unreserved and unqualified support of all members."[137] This would require a change in present practice as " . . . no one country in the area has yet declined the favor of having its production increased at the expense of another country that was in disfavor with the majors."[138] But "The wise men of all nations have always rightly maintained that 'where there is a will there is a way.'"[139]

A joint production programming plan would tailor supply to demand, allow quotas to the respective OPEC members, " . . . and at the same time allow the production of crude by companies only through a program which would predetermine each operator's fair

---

[136]Ibid., p. 54.

[137]Ibid.

[138]Ibid., p. 68.

[139]Ibid., p. 53.

share of the market."[140] Such a program " . . . is feasible and would be an effective way of strengthening the market, putting an end to the erosion of realized prices, and setting them on an upward trend."[141] This would give the companies greater profits which in turn could restore the level of postings " . . . to that which prevailed before August 1960,"[142] unify the various crude grade posted prices by upward adjustments, and produce a balance between exporting centers by adjusting Middle East posted prices to the level of Venezuela. Finally, it would allow the national oil companies to sell oil from their increased quotas without discount. This profit sharing between OPEC and the companies would be at the expense of the consumer, but "OPEC is not greedy . . .-- all it wants is an *equitable* price . . .; oil is not the only source of energy and its price could be in some way equated with those of other sources."[143]

Whatever one's opinion on this program one has to agree that its full implementation would indeed represent an iron-clad cartel which would radically alter the structure of the international oil industry. It could "equitably" increase the price of gas to the

---

[140]Ibid., p. 56.

[141]Ibid., p. 59.

[142]Ibid., p. 60.

[143]Ibid., p. 64. Italics in the original.

level of oil, the price of fuel oil to the level of coal, and the
price of products to the level of products derived from synthetic
oil, thus presenting the consuming countries with a rather hefty
import bill. By a further tightening of the screw, and never
mind about the regulations of G.A.T.T., it could allocate quotas
to the consuming countries on the basis of discriminatroy prices
reflecting the local energy situation. Furthermore, it would
effectively deprive the majors of their operational flexibility
and production manipulation weapon, thus relegating their role to
the one of relatively powerless middlemen fighting a losing battle
against the national oil companies of the producing countries.

There is just one flaw in thisprogram: it will not work.
Admitted that the possibilities ennuciated above may seem farfetched,
and that OPEC only wants an "equitable" price. Even though history
tends to show that the definition of any adjective describing a
distribution of economic and/or political power tends to follow
the power of the definer, let us assume that such a joint production
program would be limited to the stated aims. The question would then
be, would it last?

Let us disregard past historical experience, including
OPEC's (which shelved its international proration plans for lack
of agreement among its members), and assume the most optimistic
conditions for the implementation of such a program. Let us first

assume the satisfactory solution to the bane of any cartel, the problem of entry. Specifically, this would have to include the Soviet Union and present outsiders, future producing countries, and offshore areas beyond present national jurisdiction and exploitation feasibility. Having cooperated with the cartel of the majors in the thirties' the Soviet Union might well cooperate with the cartel of the producers in the seventies'. Present outsiders, e.g., Nigeria, have already been included in the calculations of the OPEC program. Future producing areas, which might or might not join the ranks of OPEC would sign concessions favorable to the majors and then swiftly legislate changes once that production had reached a satisfactory level. Finally, offshore production outside of present national jurisdiction might not occur, or if it did, then only in the framework of some international control which would have the effect of setting oil prices to OPEC's level. And just to be on the safe side, let us with a flourish close the loophole of synthetics by assuming the production costs of shale and tar sands oils to be above the production costs of coal oil.

Some of the above assumptions are clearly unrealistic and a breach in some of them would be enough to set the envisaged cartel into a tailspin. Never mind. The object of this exercise is to set a scenario with the most favorable assumptions in order to highlight the crucial flaw of such a proposed arrangement. The object of any cartel is to benefit its participants at the expense of the consumer. In view of the artificial separation between costs and prices there is always an incentive, specially for the lower cost producers, to expand output and/or to shade prices in order to increase the share of the market. The decisive aspect is then <u>who</u> <u>controls</u> <u>the</u> <u>controllers</u>.

At the national level control may be exercised by government if it choses to do so. At the international level, which by definition is characterized by a fragmentation of authority, the obvious question is one of enforcement. The majors have been partially successful in this respect, first by means of a cartel, later by the parallel interest inherent in any oligopoly. But the key to their success has been vertical integration which has brought the situation of parallel interests at all stages of the industry. Unless one were to assume that undeveloped countries were willing to integrate forwards, thus investing in industrialized countries, how would the producing governments fare in this respect, with export price as their only common interest?

> If there is nothing but a high price, or a high per-barrel guarantee to every producing government, every one would be constantly confronted with the knowledge that there was a considerable additional profit to be made by shading the price -- preferably, of course, hiding the reduction in some of the many other terms of the bargain, such as freight. Most of the governments most of the time would doubtlessly

> resist the temptation to make a mutually profitable deal with a concessionaire. But that is not enough under great excess capacity. It takes only one, either fearing that it has been left out, or resentful of an unduly low share, or badly pressured by a treasury deficit or balance-of-payments problem, to start the stampede to the exit. Unless there is a market sharing agreement, which will be adhered to, the price cannot be held no matter how many people sign.[144]

Granted from the above that governments would have a higher incentive to shade prices than the international majors, for whom crude prices are mostly a matter of inter-affiliate transfers. But surely governments must realize their long-term interests better than profit minded firms. Not quite.

> In many circumstances . . . a large firm . . . is in a position to take a longer view than are most political leaders of governments. Politicians tend, above all, to try to stay in power. . . . But as a result they give great weight to the immediate, and as a consequence a government may often take a very short-run view of its action.... . . In other words, governments may use a wider-angle lens through which to view events but the range of the lens is often shorter.[145]

Having dispatched, I hope successfully, the grand design of international proration on its strongest point we can then return to a more realistic appraisal of the evolving structure of the international oil industry. If radical changes are unlikely what then is the outlook? One can with reasonable safety assume a constant

---

[144] Adelman, The World Oil Outlook, p. 110.

[145] Penrose, op. cit., p. 267.

string of discoveries that will if not supplant, then at least
mitigate the predominance of Middle East oil on the international
market, but not sufficient to prevent the gradual erosion, stopping
short of liquidation, of the major's profits and operational control
in favor of the existing producing countries.[146]   This, of course,
does not preclude sporadic nationalizations (witness Bolivia),
boycotts and counter-boycotts, or -- for that matter -- partially
successful cartels.

More of the same? Presumably so, but with one important
difference. In contrast to the oil companies that deal mostly in
oil and money, the producing countries can offer something more
to the industrialized consuming countries within whose borders, in
the final analysis, the bargaining power of the companies rests.
A slippery, double edged offer, whose dangers are only compensated
by its potential rewards. In other words: themselves. But before
we enter the hot bed where international economics and politics
enlace we must first examine a seeming prosaic topic. Oil demand
based on energy competition does not involve the drama of competing
collectivities, but when all will be said and done, these will become
echose of senseless sound and fury. To advance on the conclusions
of the following chapter, this involves nothing less than the
displacement of oil from most of its markets by nuclear energy.

---

[146]For some variations on this theme see pp. 312-321.

CHAPTER III

OIL DEMAND

The previous chapter was based on the assumption that the supply of oil depends on its demand which, in turn, is largely dependent on the competition among various energy sources. This assumption is now analyzed and qualified.

I. INTRODUCTION

The relationship between oil and international relations is spacial and temporal. While the spacial factor is dictated by the topic, the temporal factor depends on the aim and curiosity of the observer. The aim followed in this thesis is the examination of the relationship between these two segments of reality during the time of their significant mutual interaction. While the past is a matter of historical record the future belongs to the realm of speculation. One way to determine the period of relevance of this topic is to estimate the winner of the race between the depletion of oil and the transformation of the nation-state system. While no sensible person is going to risk to estimate the time of the hypothetical demise on our planet of the system of autonomous centers of decision, some have ventured to determine the approximate time when man will exhaust his geological capital of liquid hydrocarbons. Such estimates are arrived at by comparing presumable world recoverable oil with long term estimated demand. Needless to say, in view of the uncertainties involved in the evaluation of both variables, such estimates tend to vary appreciably among authors. If, however, one adds to oil as found in nature the reserves of energy forms capable of being transformed into synthetic oil, one will find these estimates as tenuous as the predictions on human behavior.

One way to get out of this impasse is to apply to the factor of demand the criterion of political significance. Taking present oil consumption technology and, therefore, oil demand growth rate, as a constant, and comparing it with the estimated reserve in place of natural and synthetic oil, we can visualize centuries ahead where oil will continue to provide an indispensible share of man's energy needs and to fuel the passions of collectivities. A different picture arises if, under the impact of developing energy production and consumption technology -- with some assist from governments -- the demand for oil is brought to negative growth. In that case, after a period of decline, oil will be relegated to the status of a "normal" primary commodity. Thus, deenergized and depoliticized, it will become of marginal interest to the student of international relations. In view of this student's faith in man's technological

148.

progress, this alternative will be the basic assumption of this chapter.

But now a word on methodology. While there is no serious problem in obtaining raw data on past and present oil demand, and on its competition with other sources of energy, the future remains per definition a matter of educated guessing. Forecasts are either middle term (from 10 to 15 years), for which quantified estimates can be reasonably made, or long term, which require uncertain extrapolations from present trends and outright guesstimates of technological developments.

The most common method of forecasting oil demand is to estimate future energy demands and to assign a percentage of this to oil, either on a gross national product or on a sector-by-sector basis. In view of the long term investments necessary in the production of energy sources and of major energy-using equipments, one can state that middle range forecasts are usually fairly accurate, both in the detail and in the aggregate. Thus, according to one source, the 1960-1980 period will show a world demand growth rate of 4.7% for total energy and of 5.6% for oil.[1] The reasonable accuracy of this and other similar forecasts is blunted when one stretches the number of years. The uncertainties involved in the

---

[1]Commission des Communautés Europeénnes, Tendance Énergétique Mondiales. Serie Énergie No. 1 (Brussels: E.E.C., 1968), p. 20. The figure for oil does not consider the so-called Communist world.

art of long range forecasts have been aptly described by the former president of the High Authority of the European Coal and Steel Company:

> Hypotheses are being worked out which, among other things, cannot allow for the effects of technical changes, and for which, even given the definite correlation between economic development and energy consumption, their authors are still undecided whether to adopt the overall or the sector-by-sector method. . . . Anyway . . . it is practically impossible to make the figures arrived at by the overall method tally convincingly with those calculated by the sector-by-sector method. And then again it is necessary - very necessary - to take into account the competitive position among the different energy sources.[2]

Thus forewarned let us avoid the pitfalls of quantified hypotheses. This can best be done by accepting the estimates of others, and then by raising our sights to a number of years -- say 30 -- past which such estimates become quite unrealistic. This solution avoids among other things the calculation of the effects of prior heavy investments made in less efficient, non-covertible energy producing and consuming facilities. Past the stated period it has the additional virtue of giving presently foreseen technological developments a more important effect on the demand for oil

Now let us see what we have in store for the next three decades. Those who courageously wrestled with demographic estimates, and with GNP and energy growth rates for various countries at

---

[2]Piero Malavesti, Sources of Energy in Industrial Revolutions, Monograph of the European Community Information Service, p. 13.

different stages of development; who noted variations by sectors; who took into account energy reserves and their locations; who -- finally -- integrated technological developments with energy competition, seem to agree on the following. By the end of this century, the growth rate of oil demand will be smaller than the growth rates of total energy demand and of the gross national product. While their estimates differ considerably (a difference of 1% of growth rate may mean a difference of billions of barrels per year) they all agree that the demand for oil will continue to grow.[3]

This should be sufficient for our purposes. Let us note a sharp oil demand growth rate curve at the beginning of the period and a gradual leveling at the end, and proceed with a description of the teething of the next millenium which, from our present vantage point, will be blessed with fewer variables. As previously mentioned, the main consideration will be the impact of technological developments on energy production and consumption. Thus, as a counterpoint to the discussion in the first chapter on the effects of technological developments on oil supply, this section will show that other technological developments will deprive oil of its major function.

---

[3] As an example of long range forecase, see Francisco R. Parra, The Development of Petroleum Resources Under the Concession System in Non-Industrialized Countries (Geneva: OPEC, 1964), pp. 9-33.

151.

A final word. While the avowed purpose of what follows is to provide the "proof" of the displacement of oil from its major energy markets and, therefore, of its loss of political significance in international relations, the used procedure will not be too unfair. As a result it will be assumed that, taking account of inflation, oil's competition with other sources of energy will not be resolved by the mere expedient of significantly higher oil production costs and prices.

II. ENERGY PRODUCTION

Technological developments in energy production affect energy consumption mainly by reducing costs in the production and transportation of primary energy sources and in their transformation into secondary energy. Accordingly, the discussion will cover both aspects.

Primary Energy

The effects of technological developments on the production and transportation costs of primary energy sources depend to a large degree on their reserves and general characteristics. While nature's energy resources are essentially static our knowledge of them is dynamic and varies with time. The concept of reserves is basically an answer to the questions: what do we know, what can we reasonably infer, and what can we guess. Table III shows actual production figures and the latest state of our knowledge, inferences

151a.

TABLE III

WORLD ENERGY RESERVES AND PRODUCTION
(in billions of barrels of oil equivalent)

Energy	Estimated in Place	Indicated & Inferred or Unproven[a]	Measured or Proven[a]	Production in 1966[b]
Oil	10,000[c]	360	396	10.86
Oil Shale	2,000,000[d]	763	698	--
Tar Sands	1			--
Natural Gas	5,400[c]	752	194	5.08
Coal	1	31,200	2,300	
Brown Coal & Lignite	1	2,635	282	11.54
Peat	1	4188		
Hydropower	per annum 9[a]			0.62
Uranium	2/1,000,000 of earth's crust[e]	252[h]	126[h]	
Thorium	12/1,000,000 of earth's crust[e]			--
Deuterium	45×10^{12}[f]	45×10^{12}[f]	45×10^{12}[f]	--

[a] World Power Conference Survey of Energy Resources, 1968
London: World Power Conference, 1968.

[b] United Nations, World Energy Supplies, 1963-1966. U.N. Document ST/STA/SER J/11, pp. 10-11.

Table III Concluded:

ᶜU.S. Department of Interior, Geological Survey Circular 522, p. 18.

ᵈU.S. Department of Interior, Geological Survey Circular 523, p. 18.

ᵉU.S. Atomic Energy Commission, Sources of Nuclear Power (Oak Ridge, Tenn.: USAEC, 1968), p. 14.

ᶠAssuming complete fusion of deuterium at 5.6×10^{10} cal. per gram, multiplied by 4.5×10^{19} grams of deuterium estimated in oceans. U.S. Atomic Energy Commission, Controlled Nuclear Fusion (Oak Ridge, Tenn.: USAEC, 1968), p. 7.

ᵍInferred and measured.

ʰAssuming 1% of U-235 in $U_3 O_8$ and energy release from U-235 only, at present average reactor conversion efficiency of 7,000 KWH pro gram of U-235. For total energy in uranium, multiply the given figures by approximately 300.

ⁱNot available.

Conversion Factors

Energy	Tons of Oil Equivalent
7.2 barrels of oil	1.0
1 ton of coal	0.7
1 ton of lignite	0.2
1000 m³ of natural gas	0.9
1 ton of peat	0.35
1000 KWH	0.25
1 gram of U-235	1.75
1 gram deuterium	8.00

and guesses.[4] From this Table we can see at a glance that, even while allowing for increased energy demand, organic energy sources may last for centuries, and that -- with the caveat of technology -- inorganic energy will endure for aeons. Having noted the order of magnitude or total energy reserves and deduced that technology has something to develop on, let us analyze its possible effects on the various energy sources according to their major characteristics.

A. Organic Energy

The effects of technological developments on the production and transportation costs of most organic energy sources have already been discussed in the context of their conversion to synthetic oil. Let us now briefly look at the other organic energy sources. Brown coal, lignite and peat are essentially the poor relations of coal. They share with coal the same process of mining and bulk transportation but have the net disadvantage of lower calorific value. In other words, equal costs for less energy. Peat is produced in very limited quantities. The competitiveness of brown coal and lignite is restricted to their locational advantage in some markets. Their markets should be further eroded by reductions in transportation costs.

---

[4]Table III does not include the so-called new energy sources, i.e., solar, geothermal, tidal and windpower, for lack of adequate measurements. They are, however, briefly discussed in the text. No addition is made because some energies are dependent for their release on future technology and because hydropower is in function of time.

Natural gas (as opposed to manufactured gas derived from coal) consists mainly of methane and of some natural gas liquids. Natural gas is usually found together with oil though it can also be found alone. It is, therefore, a joint product of oil exploration and sometimes of oil production. When not injected into the oil well in order to increase reservoir pressure, it is transported in pipelines at a cost "twice to three times as much as oil"[5] and, since very recently, in special ships in the form of refrigerated liquified petroleum gas. It is also quite expensive to store. Natural gas, which is mainly used as an unprocessed general fuel and as a raw material for the chemical industry, has become oil's chief competitor in some areas, partly due to government intervention. This competition is bound to increase with the full-scale use of previously impossible sea transport, thus putting an end to the outrageous waste of this source of energy.[6] Nevertheless, when taken in isolation from oil production and transportation technology (including nuclear fracturing), no further improvements can be expected that would significantly increase the competitiveness of natural gas.

Summing up, we can say that organic energy sources share the common trait of mining and of high transportation costs. The,

---

[5]Hartshorn, op. cit., p. 85.

[6]The amount being flared in (Venezuela and in the Middle East), it was estimated in 1960, had been reduced to about 100 million cubic meters a day, the equivalent of 600,000 barrels of oil per day." Ibid., p. 86, my emphasis.

perhaps, most important future mining technology development -- fire flooding -- will increase the competitiveness of synthetic oil. It is difficult to assess the effects of future transportation developments on the relative competitiveness of organic energy sources without taking their respective locations and markets into account. Nevertheless, if history is a guide, reductions in transportation costs will mostly benefit oil because of its intrinsic advantages over the other organic energy sources. If future technological developments in organic energy production and transportation will mostly benefit oil, we will clearly have to look elsewhere in order to prove its demise. We thus turn our attention to inorganic energy sources.

B. Inorganic Energy

It has been previously stated that the effects of technological developments on the production and transportation costs of primary energy courses depend to a large degree on their reserves and general characteristics. It has been seen that organic energy sources share the common trait of abundant reserves and high mining and transportation costs, the last two setting definite parameters to the scope of possible technological improvements. Inorganic energy sources do not have any common traits apart from nomenclature. They will, therefore, have to be discussed separately.

Hydropower

The world reserve of hydropower is determined by the relationship between environment, technology and economics. Let us first examine the basic relationship between technology and environment.

> Hydropower potential is created by the flow of water between different elevations, or the head. The energy of the water is used to turn a waterwheel and is coverted into electricity by a generator driven by the waterwheel.[7]

Let us disregard the periodic variations of electricity demand and their effects on hydropower, and note that the main production characteristic is determined by the fact that "the flow of the stream being utilized by hydropower commonly varies over both long and short periods, reflecting rainfall and runoff patterns."[8]

> Differences between short-period fluctuations in streamflow and desired variations in plant output can be overcome through the use of a small reservoir to provide pondage for the plant; long-term regulation of flow can be provided through storage reservoirs.[9]

Two factors stand out so far, namely that hydropower needs a continuous flow of water and that, in view of natural flow variations, some form of intervention is required. Let us now

---

[7]Sam H. Schurr and Bruce C. Netschert, et al., Energy in the American Economy, 1850-1975: Its History and Prospects (Baltimore: The Johns Hopkins University Press, 1960), p. 439.

[8]Ibid., p. 440.

[9]Ibid.

156.

return to the question of hydropower reserve and examine the relationship between environment, technology and economics, as reflected on reserve potential measurement.

> The theoretical potential constitutes the environmental limit within which technology can be applied. The technical potential, in turn, is the limit of the resources that could be developed with a given limit of technology under no cost limits. And the economic potential delimits the resources that can be developed according to prevailing costs and load combinations.[10]

From Table III we can see that the theoretical potential is approximately equal to 9 billion barrels of oil per year. The technical potential is rather difficult to quantify due to conceptual difficulties.[11] No matter. Let us take technological improvements for granted, and the corresponding increase in technological potential, and examine the practical limitations within which technological improvements can realize the economic potential. In general terms, these practical limitations are transportation and conflicting claims on water.

The competitiveness of hydropower with other sources of energy depends mainly on the price of electricity delivered to the consumer. In comparison with thermal plants, hydropower plants have usually a higher capital cost but -- theoretically -- a free and inexhaustible energy source. Assuming for the same of simplicity

---

[10] Ibid., p. 442.

[11] Ibid., pp. 441-442.

that these two factors cancel each other out on the average, we are again faced with the problem of transportation to the market, albeit not in the form of bulk but of electricity. Two considerations are in order. The first, and most important, is that there is a distance past which transportation by wire is uneconomic. As a result the hydropower economic potential is restricted by the availability of markets within the economic transportation radius. One need only to think of the Amazon system in order to realize its implications. The second point is related to the first. It is that there is a distance past which it is cheaper to transport organic energy in bulk than in the form of electricity. Consequently, if the distance of cheaper bulk transportation overlaps at some point with the distance of economic hydropower transportation there may be an area of competition for existing markets, whose resolution depends on the circumstances of the individual cases, such as load and grid characteristics. It may be assumed, however, that the combination of technological improvements in extra high voltage transmission[12] and increasing markets nearer to the potential sites for hydropower plants will, on the long run increase the hydropower economic potential.

The largest limitation on economic hydropower potential may well prove to be the conflicting claims on water. It has been

---

[12]See Ibid., pp. 339-340 for a discussion on coal-fired thermal plant location in function of the relative costs in bulk vs. wire transportation. For technological improvements in extra high voltage transmission, see Ibid., pp. 470-473, as well as in Gerald Manners, The Geography of Energy (London: Hutchinson University Library, 1964), pp. 81-85.

seen that hydropower plants require a continuous flow of water and that, in view of natural flow variations, the flow must be regulated by storage reservoirs. These two requirements, flow volume and flow rate, may conflict with the requirements needed for other purposes. Conflicts over water volume arise mainly with irrigation. While the potentially conflicting claims depend on individual circumstances, there are many instances which require the allocation of priorities. Conflicts over the flow rate arise mainly with flood control and navigation over the efficient use of storage reservoirs. Last but not least, hydropower production may be limited by reasons of ecology and/or aesthetics. As a result of transportation problems and of conflicting claims on water, the percent of hydropower production has remained practically stationary in all the industrialized countries with the exception of the Soviet Union.

May one be permitted an aside. The resolution of conflicting claims on water -- and of their respective claimants -- is partly economic and partly political and, therefore, dependent on authority. The best example of a nearly optimal solution is provided by the Tennessee Valley Authority. The results may be rather different when the stream, valley, region or drainage basin encompasses two or more countries with indifferent or poor mutual relations. This can be observed in the cases of the Nile, the Indus, the Mekong and the Jordan!

Nuclear Energy

The main characteristic of nuclear energy is that it is a practically weightless form of energy which requires a high level of technology for its efficient release. Nuclear energy is released either by fission or by fusion. Both ways will be discussed.

1. Fission

Nuclear energy is analyzed in this section in the context of reserves, secondary energy production and energy consumption. The present discussion is limited to the question of reserves. Looking at Table III one might at first glance be surprised at the relatively limited proven and unproven uranium reserves. A further reading of the corresponding footnote reveals that these limited reserves should actually be multiplied by approximately 300 times. Let us now explain this apparent discrepancy.

Nuclear fission energy is released from atomic fuels which are a combination of fissionable and fertile materials. Energy can only be released from fissionable material, i.e., one that "readily undergoes nuclear fission when struck by neutrons."[13] Fissionable material is either found in nature or produced in reactors from fertile materials. The only fissionable material found in nature is Uranium-235; the fertile materials are Uranium-238

---

[13] United States Atomic Energy Commission/Division of Technical Information, Series "Understanding the Atom," Atomic Fuel (Oak Ridge, Tenn.: USAEC, 1964), p. 3.

and Thorium-232. The atomic fuel in most reactors consists of 3 or 4% of U-235 and the rest of U-238. The U-235 releases energy mainly in the form of heat and converts some of the fertile material into a new fissionable material, Plutonium-239. If the fertile material used is Th-232, the new fissionable material is U-233. The present generation of reactors, known as "converters," operate with a net loss of fissionable material. This is a cause for concern in view of the limited reserves of U-235, "an isotope of uranium constituting less than 1% (actually 0.71%) of the element as found in nature."[14] The hoped solution to the problem is found in technology.

It is believed that in the 1980's a new type of reactor, known as "breeder," will be perfected to the point of competitiveness with present reactors. In these new reactors the atomic fuel will consist either of Pl-239 and U-238, or of U-233 and Th-232. By achieving a conversion ratio in excess of 1.0, these "breeders" will release energy and gain fissionable material at the expense of the relatively cheap and abundant fertile materials.[15] This would mean that ". . . over a period of time, our resources of fertile materials are efficiently converted to fissionable materials and these in turn are efficiently converted to energy."[16]

---

[14]Ibid.

[15]It is estimated that the time that it takes for a breeder reactor to double its original inventory of fissionable material, the so-called "doubling time" will be of the order of 15 to 20 years. Ibid., p. 35.

[16]Ibid., p. 5.

The future of nuclear fission energy depends on the relationship between demand, technology and reserves. The demand for nuclear fission energy depends on its competitiveness with other sources of energy. Assuming a continuing and increasing competitiveness of nuclear fission energy, its demand will be mainly in function of the demand for electricity, which is doubling every 10 years. This sets a premium on the development of competitive breeder reactors. Even this development will not solve the potential fissionable material shortage until at least the end of this century. Barring a series of substantial uranium discoveries and assuming the persistence of the present growth of demand for electricity, there will still be a need for converter reactors.

> In fact, as long as the doubling interval of atomic power generating capacity is shorter than the doubling time of power breeders, we will need to operate converter reactors in combination with breeder reactors in an integrated network. In such a network the fissionable material produced by the converters would be used to help fill the inventory needs of new breeders.[17]

Considering the above factors, namely increased competitiveness of nuclear energy, a rapidly increasing demand for electricity, and large, but not abundant, reserves of fissionable materials, the future of nuclear fission energy seems to be clouded by the race between technology and depletion of reserves.

---

[17]Ibid., p. 35. Italics in the original.

> Once we reach the stage where in the aggregate we produce more fissionable material than we consume there will be no danger of depleting our fissionable assets. . . . But until that point is reached, careful fuel management will be needed.[18]

With the experience of historical hindsight one tends to wonder whether the fears of uranium shortage will not be as unfounded as they once were for oil.

2. Fusion

The promise of nuclear fission is dwarfed in comparison with the potential of nuclear fusion, but while the former is coming into increasing use the latter is presently only at the research phase. Nevertheless, nuclear fusion warrants a discussion in view of its potential overwhelming economic and political significance. First a technological aperçu.

As well known, nuclear energy can be released either by the fission of elements with high atomic number -- Uranium and higher -- or by the fusion of elements with the lowest possible atomic number, namely of ordinary Hydrogen and of its two known isotopes, Deuterium and Tritium. The problem in the case of nuclear fusion is to "confer sufficient energy on the nuclei to permit them to overcome the repelling forces."[19] This energy can be provided either by acceleration or by raising the temperature. While both

---

[18]Ibid., p. 36.

[19]U.S.A.E.C. (see p. 159), Controlled Nuclear Fusion (Oak Ridge, Tenn.: USAEC, 1968), p. 9.

ways have been used to achieve fusion, present technology does not permit to do so in an efficient and controlled manner. The use of accelerators is "much too wasteful of energy to be of practical value,"[20] while the scaling or slowing down of the H-bomb is impossible because it requires the use of a fission bomb as a trigger. Other means must therefore be found.

The practical problems that must be solved in order to achieve controlled nuclear fusion are staggering. In a capsule formula the problem consists of attaining high temperatures at given densities over a period of time. The required temperatures are much higher than the ones prevalent in the interior of the sun.[21] At high enough temperatures atoms are stripped of their electrons and become ionized. Such an ionized gas is referred to as plasma, the so-called "fourth state of matter." The main problem in the handling of plasma, in the attempt to confine it at the required temperatures for controlled nuclear fusion, "arises from the loss of energy by the nuclei as a result of striking the walls."[22] In view of the electrical properties of the plasma the best solution to this problem is of confining the plasma in a magnetic field.

---

[20]Ibid.

[21]Fusion is attained on the sun with only 15 million degrees Kelvin ($^{\circ}K$) at a pace "much too slow to be useful on earth" due to the sun's extremely high density of hydrogen. Ibid., pp. 4, 10.

[22]Ibid., p. 14.

Some of the "magnetic bottle" types presently used for controlled fusion research are mirror machines, astron, and toroidal confinement systems. And now to the state of the art. Research on controlled fusion began in the early 'fifties.

> The first goal was to obtain plasmas above the ideal ignition temperature, i.e., the lowest possible operating temperatures for a self-sustaining fusion reactor. . . . This goal has been achieved. Such plasmas, at the densities needed for controlled fusion, are now routinely produced. . . . The world effort then converged upon the problem of how to adequately confine the hot plasmas.[23]

For about a decade this problem seemed intractable. "This anomalously low confinement time (the infamous Bohm time!) was long not understood."[24] It was then realized that this was probably due to "specific modes of instabilities, one of the most likely candidates being the so-called drift mode of the universal micro-instabilities."[25] Corrective actions were taken and this seemingly unpassable barrier was pierced. The best result in plasma confinement time has been achieved, at the time of writing, at Tokomak-3 in the Soviet Union. There are expectations that Tokomak-10, presently under construction, will achieve plasma parameters close to the ones required for controlled fusion.

---

[23] Bernard J. Eastlund and William C. Gough, _The Fusion Torch: Closing the Cycle from Use to Reuse_ (Washington, D.C.: Division of Research, US.AEC, 1969), p. 1.

[24] Amasa S. Bishop, _Recent World Developments in Controlled Fusion_, paper presented on November 12, 1969, before the Plasma Physics Division of the American Physical Society, mimeo, p. 7.

[25] Ibid.

> (Soviet scientists hope that Tokomak-10 will achieve) an ion temperature of 3000 eV, densities of 10^{14} cm^{-3}, and confinement time of about half a second. . . . These values are seen to be not far removed from those required for a fusion reactor. . . . if T-10 were successful in achieving the above parameters, it could reasonably be said to have demonstrated the long-awaited scientific feasibility of controlled fusion. The time scale for the completion of T-10 is expected to be about four years from now.[26]

Assuming that controlled fusion can be achieved, what will be the salient characteristics of fusion power plants? There are three possible fuel cycles leading to two types of reactors: a fusion reactor using the Deuterium-Tritium fuel cycle would produce power by dynamic conversion while a reactor using either the Deuterium-Deuterium or the Deuterium-Helium-3 cycles would do so by direct conversion. For diagrams and explanations of both types of reactors see the Appendix to this chapter. Regardless of type fusion reactors will produce some remarkable environmental benefits.

1. Fusion power plants will not produce large quantities of radioactive wastes. While the internal structure of a fusion reactor will become highly radioactive, the waste products from a fusion reactor are not radioactive.

2. Fusion power plants operate at high conversion efficiencies -- of the order of 60% or higher. Therefore, they will greatly reduce problems associated with waste heat.

3. Fusion power plants are inherently very safe. Thus we may see the day when such plants may

---

[26]Ibid., p. 13.

be installed in the center of urban complexes, reducing the costs and other disadvantages associated with long-distance power transmission and perhaps making use of the waste heat for space heating and low-temperature industrial processes.

4. Fusion power plants practically eliminate the consideration of the safeguarding of weapons-grade material from diversion for subversive purposes. A plant operating on the D-^{3}He cycle would eliminate the problem altogether.[27]

Fusion power plants will not only avoid the negative side-effects on the environment produced by fission power plants; the combination of the fusion power plant with the _fusion torch_ may alleviate, perhaps even solve, the urban waste disposal problem while at the same time allowing the economic recycling of must minerals. "Briefly, the concept of the Fusion Torch involves using the ultra high temperature plasma of a thermo-nuclear reactor to convert any material to its basic elements or to produce large amounts of ultra-violet radiation."[28]

In a world where we are rapidly depleting many of our natural resources at the same time that we are unable to adequately handle our waste, this concept seems like the answer to both the conservationists' and the technologists' prayer. And such a dual answer has not seemed very likely in the past.[29]

---

[27]Glenn T. Seaborg (Chairman of the U.S. Atomic Energy Commission), Fission and Fusion - Developments and Prospects, speech at the Council for the Advancement of Science Writing, Berkeley, Calif., November 20, 1969, mimeo, p. 16.

[28]Infra. For diagrams and abbreviated explanations of both types of fusion torches see the Appendix to this chapter. For a full discussion see Eastlund and Gough, op. cit., sections I, III, & IV.

[29]Ibid., p. 18.

And now a few words on fuel reserves and costs. The reserves of deuterium are comparable to the reserves of water. One gallon of water contains one-eighth of a gram of deuterium. Its cost of extraction from water is about 4¢, its energy content is equivalent to the combustion energy of 300 gallons of gasoline.[30] This means that at 100% efficiency, 1 gram of deuterium costing about 32¢ is equal to the energy content of 8 tons of crude oil! Tritium, on the other hand, is very rare in nature and highly expensive to produce.[31] This can be accomplished by neutron bombardment of lithium -- a common mineral -- in either fission or fusion reactors. But the tritium breeding ratio in the latter is estimated to be "at least 1.3, corresponding to a doubling time of much less than a year."[32] Thus, the cost of tritium can be limited to the initial inventory and be considered as a capital cost, or, depending on the bookkeeping, even as a very profitable investment! In regards to Helium-3, it can be obtained by the radioactive decay of tritium and mixed with deuterium for the initial inventory of a fusion power plant using the Deuterium-Helium-3 fuel cycle. No additional Helium-3

---

[30] US AEC., Controlled Nuclear Fusion, op. cit., pp. 4, 6.

[31] One kg. of tritium is worth about $1M! Amasa S. Bishop, The Status and Outlook of the World Program in Controlled Fusion Research, Presentation Made Before the National Research Council of the National Academy of Sciences on March 11, 1969, mimeo, Figure 20.

[32] Ibid., p. 16.

169.

is needed after the initial inventory because a balanced system in the fuel cycle can be achieved by "small adjustments in the ratio of D to ^{3}He, thus producing the ^{3}He via D-D fusion reactions occuring in the D-^{3}He fuel itself."[33]

We have so far examined fuel costs and reserves. It has been seen that the two basic elements for fusion power plants, deuterium and lithium, are abundant in nature and relatively inexpensive to produce, while tritium and helium-3 are rare in nature and highly expensive to produce, but that this cost is limited to the initial inventory. It is obvious that fuel costs will be only a fraction of total electricity generating costs. But estimated capital costs of fusion power plants compare very favorably with even conventional power plants.

> Recently scientists and engineers at M.I.T. and the AEC's Oak Ridge National Laboratory in Tennessee estimated what it would cost to build a fusion plant with a capacity of five million kilowatts, five times as large as the biggest atomic power plant now built. They calculated the capital cost at about $120 per kilowatt, or $20 to $80 a kilowatt less than present coal and atomic power plants.[34]

This capital cost estimate is for rather large power plants; however, these are "not out of line with those projected in the year 1990 for units using other sources of power."[35] At the same

---

[33]Seaborg, op. cit., p. 17.

[34]The Wall Street Journal, December 3, 1969.

[35]Bishop, Recent World Developments . . ., op. cit., p. 18.

time it is estimated that pure-pinch thoroidal confinement systems "offer hope of developing units with relatively small power outputs."[36] One cannot at the present time predict when fusion reactors will be put into general use, but it seems unlikely to occur before the end of this century. "From economic and engineering considerations and because of fission breeder developments and commitments already made, the earliest date that any large commercial fusion system would be likely to start operations is A.D. 2000."[37] In the meantime let us examine some other sources of inorganic energy.

C. Other Energies

The other sources of inorganic energy, the so-called "new sources," are geothermal, solar, tidal and windpower. Their novelty stems neither from their age nor from their use by man, but rather from their new application as a source of electricity. Each source will be briefly characterized.

1. Geothermal

A virtually untapped yet potentially significant source of energy is found in the earth's own heat. The most vivid example of this source of energy is the volcano. Geothermal energy proper

---

[36]Bishop, Symposium on Nuclear Fusion Reactors, September 17-26, 1969, Culham, England, Conference Summary, mimeo, p. 5.

[37]D.J. Rose, Engineering Feasibility of Controlled Fusion (Cambridge, Mass.: MIT-3980-6, Nuclear Fusion 6, 1969), pp. 183-84.

requires the application of the earth's heat to underground water reservoirs. In practice this requires the coincidence of geological faults, which allow the heat of the magma to come to or near the surface, with the abundance of underground water. When this hot water reaches the surface of the earth it is transformed to steam. Depending on the steam temperature and pressure, and output regularity, it can be used either for space heating, as in Iceland, or for the production of electricity, as in Italy. The exploitation of geothermal energy is presently of marginal significance and virtually limited to geysers. The intensive exploitation in significant commercial quantities requires the location of underground hot water reservoirs and the drilling to the required depth -- technologies very similar to oil production.[27]

There is no way for the time being of estimating the reserves of geothermal energy. One may, however, speculate on its competitiveness with other sources of energy on the basis of its general characteristics. Geothermal energy combines the production features of oil with the transportation features of hydropower. The production technologies and economics of geothermal energy can be compared with the production of an inexhaustible oil reservoir from

---

[27]For a comprehensive review of the potential, and present technology and economics, of geothermal energy, see Lawrence Lessing, "Power from the Earth's Own Heat," *Fortune*, Vol. LXXIX, No. 7, June 1969.

flowing wells on secondary production. But here the similarity ceases for steam, as opposed to oil, cannot be transported without loosing its energy. At this point geothermal energy acquires the transportation features of hydropower, namely competitiveness in function of the distance to the market. We may thus establish the idealized parameters for the competitiveness of geothermal energy with oil, used as a general fuel for the production of electricity, by incorporating the variable of distance to the market. The parameters range from priceless to worthless, with economics in between.

2. Solar Energy

The problems related to the new applications of solar energy -- the source of life -- can be reduced to the questions of technology and regularity of supply. Solar energy, which arrives to our planet in the form of photons at the rate of 1400 watts per square meter,[28] can be either converted directly to electricity via solar cells or converted into usable heat by optical concentration. There is no way to forecast how future technological developments will shape the production economics of solar energy.[29] One can, however, point out to the natural limitations of regularity of supply which, at any given location on our planet, is subject to the

---

[28] USAEC (see p. 159), Direct Conversion to Energy (Oak Ridge, Tenn.: USAEC, 1968), p. 26.

[29] For extended discussions on the technologies of the new sources of energy, see: New Sources of Energy and Economic Development. Solar Energy, Wind Energy, Geothermic Energy and Thermal Energy of the Seas. United Nations Document E/2997 St/ECA 47 (New York, ECOSOC, 1957).

variations of daylight hours, season and weather. In view of the above, the future widespread use of solar energy for the production of electricity in commercial quantities will require highly flexible grid systems.

3. Tidal Power

Tidal potential is created by the flow of water caused by "the alternate rising and falling of the surface of the ocean . . . that occurs twice a day and is caused by the gravitational attraction of the sun and moon occuring unequally on different parts of the earth."[30] The energy of the water is converted to electricity by means of a technology similar to the harnessing of hydropower. In practice, the use of tidal energy is restricted to such limited sites that permit the control of significant currents, thus relegating this vast source of energy to a place of marginal importance.

4. Wind Power

The use of wind energy for the production of electricity is hampered by the irregularity -- in time, space and intensity -- of air movement. While its contribution for the production of electricity is bound to be minimal, it may well, given adequate technology, turn out to be a boon to rural, windy areas, that are not connected to power grids.

---

[30] Webter's Seventh Collegiate Dictionary.

Let us sum up the discussion on the possible effects of technological developments on the production and transportation costs of primary energy sources in light of their reserves and general characteristics. The limiting factor for the technological development of organic energy sources is their common trait of mining and bulk transportation. In the case of hydropower the main limiting factor is relatively limited reserve potential compounded by conflicting claims on water. Within these limitations, technology's future role is limited to improvements in hydropower plants and in extra high voltage transmission. The latter, together with developments in the exploration of hot water underground reservoirs may unlock the yet unknown geothermal energy reserves. The role of technological developments in the use of solar energy is unknown, yet somewhat limited by the inherent irregularity of solar energy supply. Tidal and wind power are bound to be of marginal importance. We thus arrive to weightless nuclear energy, whose potential for cheap, virtually inexhaustible energy is only limited by man's ingenuity.

Having defined the parameters within which future technological developments may affect the competition among the various primary energy sources does not yet imply their respective postures in the production of secondary energy. To this we now turn our attention.

Secondary Energy

Technological developments in the production of secondary energy may affect the competition among the various primary energy sources in two ways. The first way is by determining the primary energy source from which electricity is produced. The other way is by reducing electricity production and transportation costs, thereby affecting the competitiveness of secondary energy in other markets. Thus, in effect, we are dealing with developments in the technology of energy conversion. Figure II shows the energy conversion matrix. Reference to this matrix will be made during the discussion of secondary energy and of energy consumption.

A. Primary Energy Source

The remainder of this century will see the competition between nuclear and organic energy for the production of electricity. After that period, nuclear energy should predominate. The other primary energy sources may be discounted for various reasons. Let us now proceed with the process of elimination. During this discussion we will neglect future fusion power plants.

The first step concerns the possible effects of developments in direct energy conversion technology on primary energy competition. If we look at the energy conversion matrix we can promptly discard most of the theoretically possible direct energy conversions. Leaving gravitional energy as an unknown we can disregard direct electromagnetic, nuclear and thermal conversions to electricity, insofar as these

FIGURE II

ENERGY CONVERSION MATRIX

To ↓ \ From →	Electro-magnetic	Chemical	Nuclear	Thermal	Mechanical	Electrical	Gravitational
Electro-magnetic		Fireflies	A-bomb	Hot iron	Cyclotron	TV transmitter	Unknown
Chemical	Photosynthesis		Ionization	Boiling (water/steam)	Radiolysis	Battery charging	Unknown
Nuclear	Gamma-neutron reactions	Unknown		Unknown	Unknown	Unknown	Unknown
Thermal	Solar absorber	Combustion	Fission Fusion		Friction	Resistance heating	Unknown
Mechanical	Solar Cell	Muscle	A-bomb	Turbines Engines		Motors	Falling Objects
Electrical	Solar Cell	Fuel Cell Battery	Nuclear Battery	Thermo-electricity	Generator Turbine MHD*		Unknown
Gravitational	Unknown	Unknown	Unknown	Unknown	Rising Objects	Unknown	

*Magneto-hydro-dynamics

Source: USAEC, <u>Direct Conversion of Energy</u>, op. cit., pp. 6-7.

conversions take place in low-power energy conversion equipment, and turn out attention to direct conversions from chemical and mechanical energy.[31] In respect to the former we find a potentially significant development in the technology of the fuel cell.

> Perhaps the most challenging task contemplated for the fuel cell is to bring about the consumption of raw or slightly processed coal, gas and oil fuels with atmospheric oxygen. If fuel cells can be made to use these abundant fuels, then the high natural conversion efficiency of the fuel cells will make them economically superior to the lower efficiency steam-electric plants now in commercial service.[32]

There is no way of estimating when, if at all, this development is likely to occur, nor its actual economics. One may, however, venture a few observations. In the first place, one may presume that it will affect all organic energy sources equally. Secondly, that by the time that the fuel cell is perfected further cost reductions in nuclear energy electricity generation will have occured, thus blunting the competitive edge of the fuel cell. Thirdly, and discounting the direct use of fuel cells in the residential and industrial sectors, that with unknown capital costs, it will mainly affect the

---

[31]Direct electromagnetic and nuclear conversions to electricity occur with the use of semiconductors. Direct thermal to eletrical conversions may take the form of thermoelectricity, which uses semiconductors and thermoelectric couples, or of thermionic, ferroelectric and thermomagnetic conversions. It is difficult to determine how future technological developments will affect the conversion efficiencies and the order of magnitude of power level at which these latter three direct thermal conversions may take place, but present indications are that these will be confined to low-power specialized uses. USAEC, <u>Direct Conversion of Energy</u>,

fuel component of total electricity production costs. Finally, that improvements in conversion efficiencies do not mean linear reductions in fuel costs and that, past a certain point, the higher the efficiency the lower the impact on cost reductions.

We now come to the final direct conversion, that of mechanical to electrical energy. This affects tidal, wind and hydropower. It has been seen, however, that these primary energy sources have intrinsic limitations, thus lessening the impact that any technological improvements in conversion efficiency might bring.[33] Having exhausted the direct conversions, let us proceed with indirect or dynamic conversions, namely from thermal to mechanical. By leaving solar and geothermal energy as unknown quantities,[34] we narrow down the field

---

op. cit., pp. 12-19, 26-33. In any case such developments would presumably not affect the main competition between nuclear and organic energy sources.

[32]Ibid., p. 24.

[33]"One does not need to infer . . . that great opportunities exist here for enlarging the economic potential or for significantly accelerating the pace of hydro development. But to the extent that such a situation may prevail, the opportunities are present for higher efficiencies, hence lower costs, in future hydro installations." Schurr & Netschert, et al., op. cit., p. 467.

[34]Given an adequate distance to the market, geothermal energy has the additional advantage of low capital cost thermal station. Given adequate natural steam pressures and temperatures and the recycling of the water into the ground, geothermal stations can be practically limited to the turbogenerator and a few pumps.

to organic and nuclear energy sources. Of the organic sources we may disregard oil shale, peat, brown coal and lignite because of their low calorific value, thus relegating their competitiveness to limited geographical and political considerations. We then finally remain with coal, oil and natural gas competing with nuclear power in thermal stations.[35]

The long range role of oil in the production of electricity is most likely to become insignificant due to a combination of technological and political factors. It has been estimated that by 1980 oil's share of the U.S. electric utility market will have shrunk to 2%.[36] This may well be indicative of oil's long range prospects in the world production of electricity.[37]

---

[35]This leaves to oil the small markets unconnected to power grids, as "Market size also influences the means by which electricity is generated, for it is felt that for units with a capacity of less than about 250KW the advantages of diesel as opposed to steam generation are overwhelming. Manners, op. cit., p. 119.

[36]The Chase Manhattan Bank, Outlook for Energy in the United States (New York: The Chase Manhattan Bank, 1968), p. 29.

[37]While the long range competition between conventional and nuclear energy seems to be a foregone conclusion, oil is further handicapped by government policies in industrialized countries. This is discussed in greater detail in the next chapter. At this point it may be mentioned that O.E.C.D. countries will presumably continue to protect their domestic energy sources at the expense of cheap oil imports, and that developing nations will continue to import oil until such time when nuclear energy becomes fully competitive in their markets.

B. Production Costs

Let us accept the hypothesis that by the year 2000 the U.S. electric utilities market will be shared equally between conventional and nuclear energy,[38] and assume that this projection is representative for other industrial countries. Let us further assume that the introduction of competitive breeder reactors in the 1980's will solve the long range world problem of fissionable material reserves. The ultimate projection of this trend would be the use of nuclear energy for grid base load and of conventional thermal and other stations for peak electricity requirements. Competitiveness between energy sources for the production of electricity does not, however, given an indication of actual production costs. These are now analyzed from the standpoint of technological limitations on possible cost reductions.

The first point to be considered is whether thermal plants, regardless of type, have inherent limitations within which technology can operate. This brings us to the concept of thermal efficiency. Let us first establish the theoretical parameters and consider the first two laws of thermodynamics. The first, the Law of Conservation of Energy and Mass, which Einstein related in his $E = mc^2$ formula, shows that energy cannot appear without the disappearance of mass.

---

[38]USAEC, Civilian Nuclear Power - A Report to the President, 1962, p. 43, as quoted in USAEC, Atomic Fuel, op. cit., pp. 35-36.

The Second Law shows that heat cannot be transformed into another form of energy with 100% efficiency. Within the confines of these two laws, which have been paraphrased, respectively, as (1) you can't win; (2) you can't even break even, efficiency is further restricted by more earthly considerations, namely our planet's temperature. The only practical way to overcome this handicap is to operate at as high as possible temperatures, even if this involves an uphill struggle against an exponential curve.[39]

---

[39]The problem of low heat conversion efficiency is explained by Carnot's efficiency formula for heat engines:

$$e = 1 - \frac{T_c}{T_h} = \frac{T_h - T_c}{T_h}$$

where: e = the so-called Carnot efficiency
 T_c = the temperature of the waste heat reservoir (in °K)
 T_h = the temperature of the heat source (in °K)

Since the Kelvin temperature scale starts with zero at absolute zero, instead of at the freezing point of water, and since T_c -- whether atmosphere of water -- has an average temperature of 300 °K, the most practical solution to improve the efficiency is to raise T_h as high as possible without melting the engine.
Paraphrased from USAEC, Direct Conversion of Energy, op. cit., p. 11.

From the above formula we see that e is related to T_c and T_h, by a fraction, therefore exponentially. In view of our planet surface temperature, the thermal efficiency temperatures range between 300 and 30,000 °K.

TABLE IV

THERMAL EFFICIENCY IN FUNCTION OF STEAM TEMPERATURE

Year	Pounds Coal/ KWH[a]	Percent Efficiency	T_h in $°K$[b]
Theoretical minimum	28	1	
1900	6.85[c]	4	<373[e]
1920	3[c]	9	
1955	0.95[c]	28	420
1965	0.7[d]	40	510
1985?	0.42	60	900[f]
Theoretical maximum	0.279	99.9	>30,000

[a]Coal of 12,263 Btu as used by U.S. utilities (Schurr & Netschert, et al., op. cit., p. 184), needed to produce a KWH (3,412 Btu).

[b]T_h in Carnot's efficiency formula, derived from efficiency column, without discount of thermal plant energy consumption, using present technological level as a constant.

[c]Average for the United States during the stated years (Schurr & Netschert, et al., op. cit., pp. 180-81).

[d]Conversion efficiency for modern plants only. (USAEC, Atomic Fuel, op. cit., p. 2).

[e]Boiling temperature of water.

[f]Assumed heat attained with magneto-hydro-dynamic generators with corresponding efficiencies and coal consumption.

Now let us see how these theoretical considerations affect thermal plant conversion efficiencies and therefore, part of the fuel component of total electricity generating costs. Table IV is part historical and part theoretical. It shows the amount of coal necessary to produce a KWH and the resulting conversion efficiencies for various years, which in turn are related to the theoretically required temperatures of the heat source taking present technological level as a constant.[40] In this table one can see the characteristics of the exponential curve. At very low conversion efficiency levels it takes minimal increases in temperature to achieve large reductions of fuel consumption pro KWH produced. This reverses as one improves the efficiency until a point where it takes extremely high temperature increases to make minute changes in the consumption of fuel.

Small as these reductions might be pro KWH, they acquire a vast importance in view of the order of magnitude of needed electricity.[41]

---

[40]The relation between conversion efficiency and steam temperature is an obvious oversimplification which neglects on one hand all the power losses directly related to thermal plant operations, and all past technological developments, on the other. It does, however, show the conversion efficiency limitations within which future technology must operate.

[41]It is estimated that by the year 2000 the U.S. will produce 8.6 trillion KWH from fuel burning plants. <u>Bituminous Coal Facts, 1968</u>, op. cit., p. 17. A 20% increase in efficiency, from say 40 to 60% would reduce the amount of coal needed to produce a KWH by 0.28 and thus produce savings of roughly one billion tons of coal equivalent per year. This figure shows the order of magnitude of savings in conventional plants only. Different calculations must be made for nuclear plants because of their particular fuel use characteristics.

Thus, any future increase in thermal plant operating temperature will somewhat decrease electricity production costs, while at the same time effecting large savings of primary energy to the economy as a whole. It is hoped that sometimes in the 1980's the magneto-hydro-dynamic generator will be perfected.[42] By substituting the conventional rotating turbogenerator for a static duct with plasma it holds prospects of increasing the temperature in thermal stations by hundreds of degrees and, therefore, of increasing the fuel conversion efficiency.[43]

Having established the limits for thermal plant conversion efficiency within which technology can have a significant effect, let us examine the possible impact of technological developments on electricity production costs for conventional and nuclear thermal plants. Electricity production costs can be broken down according to capital charges, fuel costs, and operations and maintenance.

A very rough idea of the present difference in cost structure for large stations may be obtained from the table shown below, assuming coal to be priced at about $0.25 per million BTU:[44]

Cost	Percent of Total	
	Coal	Nuclear
Capital charges	46	57
Fuel costs	48	31
Operation & maintenance, etc.	6	12
	100	100

---

[42]Magneto-hydro-dynamic generation may perhaps be perfected sooner, " . . . but its contribution up to 1980 is likely to be negligible." Organization for Economic Cooperation and Development, *Energy Policy, Problems and Objectives* (Paris: OECD, 1966), p. 168.

186.

First a look at conventional thermal stations. While higher fuel conversion efficiencies will affect conventional and nuclear stations alike, few, if any, reductions can be expected in the cost of organic fuel delivered to the conventional thermal station. We can neglect operations and maintenance costs due to their small impact on total costs. In regards to capital charges, conventional thermal stations have reached a technological plateau where, "further reductions in the capital costs of conventional stations are expected, though at a slower rate than in the past."[45] The only significant improvement in capital charges that can be expected is in the increase of thermal station capacity. In this respect, however, conventional stations are at a competitive disadvantage with nuclear stations, both in cost reductions achieved at every increase in station capacity and, in the long run more important, in the absolute station capacity past which cost reductions may be significant.[46]

---

[43]USAEC, Direct Conversion of Energy, op. cit., pp. 11-12, 19-20.

[44]Parra, op. cit., p. 28.

[45]OECD, Energy Policy, op. cit., p. 59.

[46]"One can as of now foresee cost reductions by the use of more powerful units. This possibility is already important for conventional stations. By extrapolating from the present situation one can foresee that by successive passage from the 300 MW to the 600 MW stage, then from the 600 MW to the 900 MW stage, installation costs

187.

A different approach must be used for the analysis of nuclear stations, "because in general high capital cost types of reactors have low fuel costs and vice versa."⁴⁷ Nuclear reactors are classified according to the material which carries away the heat from the core. The reactors so far in commission are cooled by either boiling or pressurized light water, gas, heavy water, and liquid metal. Experience so far has shown that the material of the cooling system determines the power level at which the reactor can operate,⁴⁸ the richness of the fuel that can be used, and the conversion factor of production of new fissionable material. The richness of the fuel (i.e., the percentage of fissionable material), determines initial fuel cost and influences capital costs.⁴⁹ Stated in an oversimplified capsule formula, the

---

for coal and fuel oil will decrease, respectively, by 13 and 14%, and by 8 and 9%. For nuclear stations, capital cost reductions of 25 and 15% respectively, can be achieved by the passage of the same stages. Past 900 MW a significant decrease can only be foreseen for nuclear stations." Commission des Communautés Europeénnes, Première Orientation pour une Politique Énergétique Communautaire. (Communication de la Commission au Conseil). (Com (68) 1040), Annexes I et II. Brussels, January 17 1969, pp. 106-107. My translation and emphasis.

⁴⁷OECD, Energy Policy, op. cit., p. 59.

⁴⁸"One of the most interesting things about nuclear reactors is that they are capable, in principle, of operating at virtually any power level; the limiting factor, from a practical standpoint, is the rate at which the cooling system can carry the heat away from the core." USAEC (see p. 159), Nuclear Reactors (Oak Ridge, Tenn.: USAEC, 1967), p. 11.

⁴⁹The more efficient the cooling system, the poorer the fuel that can be used, and therefore the lower the fuel cost. On the other hand, "the richer the fuel is in fissionable atoms, the more compact the reactor." Ibid., p. 5.

efficiency of the cooling system material is in direct proportion with power level, uranium conversion efficiency and capital cost, and in inverse proportion to total fuel costs.

Thus, present reactors show different trade-offs within similar total cost range.[50] This does not mean, however, that technology must be bound within rigid parameters where the desired variable, such as total costs or conversion efficiency, must be offset at the expense of the others. The relation between the various factors enumerated above is not one of rigid mathematical proportion, but rather one of trend and influence. Improvements can, have, and will be made in many of the numerous components that comprise the reactor design and fuel cycle.[51] After all, "atomic power is just beginning to emerge from the cocoon of research and development."[52] One could

---

[50] The light water boiling and pressurized reactors operate at 547 and 588 °K, respectively, with a high net consumption of enriched uranium, but with relatively low capital costs. Gas-cooled, graphite moderated reactors operate at 811 °K, have a lower net consumption of fuel -- natural uranium -- but have high capital costs. The heavy water reactors operate in the middle range of all these factors. The heavy water reactors are on the whole more competitive than the former three, and are seen as an interim solution until such time as the competitive breeder reactor is perfected. Liquid-metal-cooled reactors, which are actually capable of breeding, operate at 700 °K, with negative net fuel consumption, but at very high capital costs. For stated temperatures and additional data, see USAEC (see p. 159), Nuclear Power Plants (Oak Ridge, Tenn.: USAEC, 1968), pp. 5-14.

[51] As an example of the above, cost reductions are expected in the fabrication of already enriched fuel elements, which are presently the "largest single factor in the cost of atomic fuel." USAEC, Atomic Fuel, op. cit., pp. 17-18.

[52] USAEC, Nuclear Reactors, op. cit., p. 20.

say that the inverse proportion between fuel and capital costs is not so much a static situation where technology must operate within natural constraints, but rather a dynamic one where improvements in one direction spur advances in another.

> The performance of any reactor depends on the performance limits of its basic materials. Research on reactor materials, notably fuels, and on other reactor components (pumps, valves, etc.) is constantly creating new design possibilities. Therefore the relative merits of alternative designs requires frequent re-evaluation. This is healthy as it inevitably stimulates renewed development efforts.[53]

Summing up, we see that within the limitations set for all heat engines, nuclear energy poses great challenges but few limitations to technological developments. These in turn may significantly reduce electricity production costs by qualitative improvements in thermal stations and by a large increase in their capacity. The problem of integrating future large power units of over 1000 MW into power grids is partly technical but mostly economic in nature.[54] Nuclear power has progressed from an idea to the drawing board and is now entering the initial phase of large scale production.[55] While there must be a limit past which technological developments bring only marginal results, this limit is clearly not in present sight.

---

[53]Ibid., p. 13.

[54]The economic aspect is largely a matter of distribution of base and point load among existing power units according to demand volume and density. The full impact of large units with lower pro KWH cost will be felt as demand increases in volume and density and as less efficient units are progressively phased out. This characteristic favors urban over rural areas and industrialized over developing countries.

III. ENERGY CONSUMPTION

At any point in time energy competition takes place within the framework of energy availability, and production and consumption technology. The first two points have already been discussed; we can now turn our attention to the latter. Oil derives a large degree of its competitiveness from its unique versatility which allows it to compete with other sources of energy in most markets while maintaining a near monopoly in the field of transportation. If we are to depoliticize oil we must provide other sources of energy -- mainly nuclear -- with the same versatility at comparable price and efficiency. But first a glance into the past.

Figure III shows the main energy sources and uses according to industrial revolutions. As any capsule description of a vast multi-causal development, this Figure is only a rough approximation of the described phenomenon. In view of the different developments among industries, sectors and countries, the various entries correspond approximately to their predominent use in the urban areas of the main industrial countries. While it is fully realized that the stone and

---

[55] It has been estimated that ". . . nuclear power would be competitive on base load in the early 1970's with stations using fuel costing about $13 per ton oil equivalent and by the late 1970's with stations using fuel costing about $10 per ton." OECD, Energy Policy, op. cit., p. 60.

191.

FIGURE III

MAIN ENERGY USE AND SOURCE ACCORDING TO INDUSTRIAL REVOLUTIONS

Function		First Industrial Revolution (1750-1900)		Second Industrial Revolution (1900-1950)		Third Industrial Revolution (1950- ?)	
		Agent	Energy	Agent	Energy	Agent	Energy
a	Temperature	Range Chimney	Wood, Coal Coal	Range Boiler Chimney	Gas Oil, Coal Wood, Coal	Range Boiler Air control	Gas, Elec. Gas, Oil Electricity
	Work	Muscles	Man	Appliances	Electricity	Appliances	Electricity
	Light	Lamp	Various oils. Coal, gas	Light bulb	Electricity	Light bulb and tube	Electricity
b	Process heat	Boiler	Wood, Coal	Boiler	Coal, Oil	Boiler	Gas, Oil, Coal Nuclear
	Work	Muscle Mill Steam engine	Man Water, Wind Wood, Coal	Steam engine Electricity	Coal Coal & hydro-power	Electricity	Gas, Oil, Coal Hydropower Nuclear
c	Land Road	Muscle	Animal	Internal combustion engine	Oil	Internal combustion engine	Oil
	Land Rail	Steam engine locomotive	Wood, Coal	Steam engine locomotive	Coal, Oil	Diesel & Electric locomotives	Oil Electricity

Figure III concluded:

Function	First Industrial Revolution (1750-1900)		Second Industrial Revolution (1900-1950)		Third Industrial Revolution (1950 - ?)	
	Agent	Energy	Agent	Energy	Agent	Energy
Water	Sail Steam engine	Wind Wood, Coal	Steam engine Internal combustion	Coal, Oil Oil	Steam turbine Internal combustion	Oil, Nuclear Oil
Air			Internal combustion engine	Oil	Internal combustion & jet engines	Oil
Space					Rockets	Liquid Solid Nuclear

¹a - residence; b - industry and commerce; c - transportation.

space ages may well coexist on our planet, this Figure should give an idea of the main trends. It could be useful at some points of this discussion to refer back to Figure II for some of the energy conversions.

Let us first throw a glimpse at pre-industrial society which, by definition, is characterized by the absence of engines. Its hallmarks are animate power (muscle), water and wind mills for the production of work, sailing vessels for transportation, wood for heat, and vegetable and animal oils for illumination.

The first industrial revolution was the age of coal, whose combustion fueled the blast furnace and the steam engine. The high calorific content of coal allowed a quantum leap in the field of metallurgy, specially in the production of large quantities of pig iron and steel. The abundance of coal and iron ore allowed the large scale exploitation of the steam engine with its drastic effects on industrial production and transportation. At the end of that period, the role of the newly discovered oil was limited mainly to lubrication and illumination.[56]

The second industrial revolution corresponds approximately to the first half of the twentieth century. It is the age of chemistry,

---

[56] The United States was an exception, insofar as its steam engine fuel was wood, mainly in the case of railroads. In fact it was only as late as 1885 that the consumption of coal overtook the consumption of wood. Schurr, Netschert, et al., op. cit., pp. 60-61, 67-68.

electricity and the internal combustion enegine. Electricity replaces oil in its function of illumination, but in turn, it replaces or competes with coal as a general fuel. In addition to its competition with coal, oil demand grows rapidly with the increasing use of the internal combustion engine, whether on land, sea or air.

The third and present industrial revolution is highlighted by nuclear energy, electronics and automation. In residential use coal has been all but replaced by electricity, oil and gas. With few exceptions, the same has occured for the production of process heat. The steam engine has been abandoned as an instrument of direct industrial use and replaced by electricity, which in turn is produced from various competing sources. In the area of transportation oil is intimately bound with the internal combustion and thermal expansion engines. Railroads are partially electrified; nuclear propulsion is so far confined to military ships. With compensations for the further replacement of coal, the demand for oil grows approximately at pace with the growth of the economy.

If we peer into the future, into say a post-industrial society, we can safely assume that it will not only be a quantitative extension of present trends, but that it shall also produce its inevitable proportion of qualitative changes with direct repercussions on the demand for oil. From an energy point of view one could characterize a post-industrial society as one that satisfies most of its energy needs from inorganic energy sources. In such a society the technology of energy production and consumption could well dispense with oil. Technology must make such a development possible and, assuming energy competition without undue government interference, economic as well. Before estimating the approximate arrival of this utopia let us examine the main categories of energy sectors and/or energy-using equipments.

Residential Sector

In the residential sector, which is here interpreted as individual dwellings, the governing factor in energy competition is not so much technology as economics. Leaving all but a few appliances as a monopoly of electricity, the competition between the various sources of energy for the production of residential heat can be reduced to the consumer's balance between usually cheaper oil and gas, and more convenient electricity.[57] The role of technological developments seems to be limited to the production of low cost electricity and to the better thermal insulation of houses.[58] It can safely be presumed that

---

[57] While the respective oil, gas and electricity costs vary largely among regions and countries, electricity must operate under the handicap of low conversion efficiency. "The consumption of oil or gas on site in modern burners will yield up to 80 percent efficiency. But, because of losses incurred in generating and transmission, the same fuels will yield on the average no more than 30 percent efficiency when converted to electricity. The Chase Manhattan Bank, op. cit., p. 32.

a substantial decrease in electricity prices would crowd oil and gas (let alone coal) out of the residential market. A somewhat less sanguine assumption is that "with the improvement of the standard of living the consumer attaches less importance to the factor of cost and increasing importance to such factors as handling ease and comfort."[59]

Industrial Sector

Domestic comfort and convenience may well be worth the slight cost increase incurred with electrical heating, but for the larger heat requirements of multiple dwellings, commerce and many phases of industry, the balance reverts to organic fuels, thus achieving the happy solution of personal (corporate or state) savings at the expense of someone else's inconvenience. A feasible but as yet uneconomic alternative to organic fuels for the production of low-temperature steam can be found in the use of nuclear energy.

> Use of reactors for low-temperature (up to 400 °F), low pressure steam for use in common manufacturing operations (drying, evaporation, distillation, etc.) or for ordinary building heating also has been studied. . . . The studies indicate that it will be some time before process heat reactors can be built and operated cheaply enough to substitute for ordinary low-pressure steam boilers.[60]

---

[58]Improvements in housing thermal insulation may affect oil demand either directly, by reducing pro house oil demand, or indirectly, by lowering heating costs, thereby encouraging the heating by electricity.

[59]EEC Commission, COM (68) 1040, op. cit., p. 12. My translation.

[60]USAEC, Nuclear Reactors, op. cit., p. 24.

The problems related to the production of low-temperature steam from reactors for mutliple dwellings, commerce and urban industries, are much more complex than the ones encountered in the production of electricity, because steam -- as many good wines -- does not travel well. This would first require the location of fission reactors in urban centers, a somewhat dubitable proposition in areas where the population has a voice in the management of its own affairs. Assuming the solution to this problem, one would next be faced with the fragmented structure of urban demand. One could technically conceive a "steam grid" within a limited radius with its own base and point loads, but its economics would be anybody's guess. It would then seem that the solution to this problem would require a series of perhaps integrated, compact reactors, using highly enriched -- and therefore expensive -- atomic fuel.

While the competition with organic energy sources for the production of low-temperature steam in urban areas is dependent on future reactor technology, a lively competition for the production of process heat outside of urban areas can already be provided with present reactor technology. The theory is quite simple: high-temperature, high pressure steam for the production of electricity, followed by low-temperature, low pressure steam for the requirements of industry. The economic problem of such a dual purpose plant is that cheap electricity requires economies of scale while the demand for process heat usually does not. The solution to this economic impasse can be sought in improvements in reactor technology and, perhaps, in the clustering around the reactor of industrial plants with heavy demand for process heat. At the present time the only offsprint of this marriage between technology and economics is the nuclear sea water desalting plant. To which one might add that in the future this mining of the sea may also produce mineral by-products, thus transforming this marriage into a three-cornered affair.

> This leads to the possibility that eventually there may be large <u>tri-purpose plants</u> incorporating a nuclear power plant to produce large quantities of electricity, millions of gallons of potable water, and valuable mineral by-products.61

When one considers the present and future increasing world demand for fresh water compared with its static supply, one can see that this is a potentially vast growth market from which oil should be largely excluded.62 Even by neglecting other nuclear energy industrial uses, such as "the high temperature applications, in the range of 1500 to 3000 °F, for certain chemical processes, including the gasification of coal . . ."63 one still arrives at mixed conclusions in regards to the energy competition for the

---

61USAEC (see p. 159), <u>Nuclear Energy for Desalting</u> (Oak Ridge, Tenn.: USAEC, 1967), pp. 25-26. Italics in the original.

62"No detailed study of worldwide water requirements exists, but in 1964 the United Nations published the results of a survey of 43 countries entitled <u>Water Desalination in Developing Countries</u>. It found that there are about 20 areas of the world that need desalting and 41 others that may soon." <u>Ibid.</u>, p. 39.

63USAEC, <u>Nuclear Reactors</u>, <u>op. cit.</u>, p. 24. For further discussion on coal gasification, see also Schurr, Netschert, <u>et al.</u>, <u>op. cit.</u>, pp. 42-21.

production of low-temperature steam. While developments in reactor technology may make large inroads in oil's general fuel markets, their impact until the development of fusion reactors will presumably be abated by economic and safety factors.

Transportation Sector

So far we have seen the possible energy competition in three sectors, namely electric utilities, residential, and commercial and industrial. In the above, fuel oil and other organic energy sources used in boilers can be replaced by nuclear energy, either directly or by electricity, provided that developments in reactor technology make this feasible and economical. A different situation exists in the transportation sector, which represents over half of oil's market, due to a serious imbalance in present transportation energy-using equipment technology.[64]

> The transportation market differs from the other three major categories in one significant respect; its energy needs are satisfied almost entirely by a single primary energy source . . . There is, therefore, virtually no inter-energy competition, nor are there practical alternatives. And no change in the situation is foreseen by 1980. The petroleum products do their work effectively and at low cost. To displace them, another energy source would need to perform better at equal cost, or as well at lower cost.[65]

---

[64] It is estimated that the 1980 United States oil breakdown by markets will be: 60.5% transportation; 26.8% industry and commerce; 9.9% residential; and 2.8% electric utilities. The Chase Manhattan Bank, op. cit., p. 43, Figure 30. While these figures are not necessarily representative of other countries, they show the order of magnitude of the transportation sector.

[65] Ibid., p. 35.

The above quotation adequately reflects the role of oil in the
transportation sector for the near future. One might perhaps qualify
the concept of cost according to its social, political, and even ecological
components, but one is forced to bow to evidence: with present technology,
oil has an effective stranglehold on the vital transportation sector. Let
us, therefore, assess the possibilities of future transportation technology,
and the possible impact that nuclear energy may have in reducing oil's
monopoly in that sector.

The basic characteristic of nuclear energy is that it "combines
a uniquely mobile fuel source and a cumbersome and costly apparatus in
which to obtain and use it."[66] Thus, unless we enter the fertile field of
science-fiction, and even while allowing for considerable improvements in
reactor technology, we will still accept this basic limitation as a given.
So, without further ado, let us start with an assault on the so far
impregnable internal combustion automobile engine. But first a few words
on its history.

A. Motor Vehicles

    Gas engines, the forerunner of all present-day
combustion engines, had been developed in principle
as early as the seventeenth century. That no results
of practical value were achieved for about 200 years
was largely due to the lack of a suitable fuel.
. . . About 1870 - a decade after the birth of the oil
industry - the new fuel, gasoline, began to be used
experimentally; and, in 1876, the first gasoline-fueled,
four-stroke cycle engine, in which the gas is compressed
before ignition, was constructed in Germany. . . . Ten

---

[66]Schurr, Netschert, et al., op. cit., p. 24.

years after this invention, the first Benz motor car
was patented. . . . During the following few years,
the most important features of present-day automobile
engines were added, and in the early 1890's motor cars
were developed which proved so efficient and successful
that to the present day there have been no fundamental
changes in the basic principles of the ordinary automobile engine.[67]

Let us disregard the historical importance of the motor vehicle, and its social, economic, and political impact on the twentieth century. We will also neglect the present and future role of the automobile on the quality of urban life, as the subject is too painful. The basic drawback of the internal combustion engine is that it operates at low efficiency and that it spews hydrocarbons, carbon monoxide and nitrogen oxide in the atmosphere.[68] Leaving indignation over air pollution aside, let us examine the effect of low efficiency on oil demand. First a few figures to fix ideas. The present oil product breakdown of the U.S. transportation sector is as follows:[69]

        75% gasoline, of which 66% is for passenger cars.
        10% turbine fuel for jet aircraft.
        10% diesel for trucks, buses, locomotives and ships.
        5% heavy fuel oil for ships.

---

[67] Ibid., p. 115.

[68] "The basic facts are that automobiles now account for over 60 per cent of all air pollution in this country, while in urban areas that figure goes up to 85 per cent. A recent Senate Commerce Committee report estimated that cars annually dump over 90 million tons of pollutants into the air Americans breathe; it noted that present approaches - which focus on limiting certain types of exhaust emission - may permit a doubling of pollution levels within thirty years as the number of cars increases." New York Times, September 10, 1969, Editorial.

[69] The Chase Manhattan Bank, op. cit., p. 37.

Considering that the gasoline-fueled internal combustion engine operates at approximately 25% efficiency, that it represents 75% of the transportation market, which in turn will be 60.5% of the total U.S. 1980 oil demand, one can by simple arithmetic see the impact of the low internal combustion engine efficiency on total U.S. oil demand. Having noted the importance of efficiency, let us examine the possible effect of improvements in passenger car engines on oil demand.

Reverting to Figure II one notes two practical energy conversions, namely thermal and electrical to mechanical energy. There are two practical types of heat engines, the air-gas and the steam engines. While many improvements may be expected in the former type, which can be either internal combustion or gasoline turbine, it seems doubtful that it can produce a significant increase in efficiency. We may thus return to a modern version of the two century old steam carriage.[70] It is difficult to guess what the efficiency of future steam engine automobiles might be, because efficiency depends on engineering designs, but if present efficiency attained with boilers in other energy-using equipment is a guide, it might well result in significant efficiency improvements.

We can now turn our attention to electrical cars which, by virtue of their escape from the heat engine straightjacket, can convert energy with next to perfect efficiency, and -- more important -- which may provide for effective energy substitution. There are two possible variations of

---

[70]"Steam carriages were as old as the Watt steam engine, but public prejudice against horseless vehicles kept them off the roads for more than a century." Schurr, Netschert, et al., op. cit., p. 115.

electrical cars: fuel cell and battery. At the present time the fuel cell is used mostly for low-power requirements, such as those encountered in the space program. The fuel cell burns hydrogen and oxygen to form water and produces electricity; but it will be recalled that it may be technically feasible to use organic fuels and atmospheric oxygen instead. While research on the fuel cell-powered automobile is carried on, it would be rash to speculate on its eventual performance and economics. One may, however, make some observations on the battery run automobile, whose entry into the market is mainly hampered by the battery operating and recharging time.[71] In regards to the latter, a newly developed technique which reduces the battery recharging time from twelve to fifteen hours, to ten to fifteen minutes, promises to remove this major stumbling block.[72] The perfection of the electric battery car will not only improve the quality of the air but replace oil with electricity, which will be increasingly produced from nuclear energy.

Having discussed the possible impact of future technological developments in motor vehicles using gasoline-fueled internal combustion engines on oil demand, we can now turn our attention to the diesel engine. But first a brief description.

---

[71]"To this day, the 'refueling' problem is one of the major obstacles holding up production of a commercially competitive electrical car." Time, August 15, 1969, p. 73.

[72]Ibid.

In 1892, the design principles of the diesel engine were patented in Germany. The basic feature of the engine is the high compression of air within the cylinder and its consequent heating to such a degree that it spontaneously ignites fuel which is injected into the compressed air. It differs from the spark-ignition gasoline engine mainly in the nature of the fuel charge. In the latter, fuel and air are mixed in a definite proportion, while in the diesel engine the air-to-fuel ratio is varied with load conditions. The high air-to-fuel ratio and the high compression thus makes the diesel engine not only more efficient than the gasoline engine in its use of fuel, but permits the use of less costly fuel.[73]

As previously stated, the diesel engine is used in the transportation sector for the propulsion of buses, trucks, locomotives and ships. We will omit the discussion on trucks and buses for the same reason we neglected the gasoline-fueled agricultural vehicles· relative unimportance in the transportation sector -- and pass directly to locomotives and ships.

B. Railroads

The choice of locomotive fuel is dictated partly by technology and partly by economics. Locomotive technology has progressed from wood and coal to diesel and electricity, and may further advance to nuclear energy, thereby providing an interesting case of energy substitution. Should the nuclear-powered locomotive replace the diesel engine locomotive, it would be a case of direct energy substitution, thus depriving oil of most of the railroad market. If on the other hand the nuclear-powered locomotive were to replace the electric locomotive, it would replace the primary energy source of the electricity used to service the railroad system. Should

---

[73]Schurr, Netscher, et al., op. cit., p. 120.

part or all of the replaced primary energy source be nuclear, it would present a situation parallel to the one encountered in the U.S. after the change over to diesel locomotives, where diesel oil replaced not only coal, but residual fuel as well.[74]

Barring the competitiveness of the nuclear-powered locomotive there still remains the competition between diesel engine and electrical locomotives. While the latter enjoys the technical advantages of higher speed, thus making it ideal for the mass transit system, its competitiveness with the former depends not so much on the relative costs of oil and electricity, but rather on the characteristics of the individual railroad systems, such as distance, intensity of use, etc. However this be, one can surmise that the growing density of urban centers and areas will create an increasing need for high-speed mass transit systems which, together with a decrease in electricity costs, should provide the incentive for additional railroad electrification. This should not only deprive oil of a growing market but also decrease the corresponding oil demand in passenger cars and aircraft.

C. Vessels

In contrast to railroads, where the impact of technological improvements is limited by the scope of their operational milieu, future technology in sea transportation may alter some patterns of world commerce. Let us first examine the competition between conventional and nuclear-powered ships.

---

[74]See *Ibid.*, for a discussion of energy substitution in the U.S. railroads after the introduction of the diesel locomotive.

The potential economic advantage of nuclear propulsion for commercial vessels are: (1) elimination of fuel tanks (oil) or bins (coal), making more space and tonnage available for cargo; and (2) improved ship utilization, due to higher cruising speed and elimination of the need for frequent refueling. At present these advantages are cancelled out by the fact that the capital costs of nuclear propulsion equipment are substantially higher than those of conventional diesel equipment. Opinion varies on when the balance will shift in favor of nuclear propulsion but it is expected that this will occur in bulk cargo applications, such as ore carriers and oil tankers, before it does in passenger or passenger-cargo service.[75]

We can safely assume that in a few decades reactor technology will substantially lower the capital costs of nuclear-powered vessels to a point where the inherent advantages of nuclear propulsion will be felt in most high-sea going vessels, thus replacing oil from a substantial and politically sensitive market. But the change-over to nuclear propulsion will not only substitute energy but also produce qualitative changes in world shipping. This merits a brief aside.

The history of vessels in one of gradual progress from the galley to the clipper, followed by a rapid succession of technological changes. The nineteenth century saw the replacement of the sailing ship by the coal-fired steam engine, which in turn was replaced by the oil-fired steamship in the twentieth century. The present competition is mainly between the fuel oil-fired steam turbine engine and the diesel-electric generator. "Today the step from oil to nuclear fuel is as great an experiment as was that from sail to steam."[76] Allowing for the usual time lag between cost-

---

[75] USAEC, Nuclear Reactors, op. cit., pp. 26-27.

[76] USAEC (see p. 159), Nuclear Power and Merchant Shipping (Oak Ridge, Tenn.: USAEC, 1965), p. 3.

neglecting military and cost-conscious civilian technology, nuclear powered commercial vessels will produce changes in the pattern of world trade, by adding new types of cargoes, new ports of call, and new shipping lanes. A brief explanation follows.

"The main benefit expected from nuclear power in maritime use is sustained long-range high-speed operations."[77] Higher speeds "can make new types of cargo economically feasible for movement by sea."[78] Such new types of cargo could well include perishable agricultural products, thus adding a new dimension to world consumer benefits and world producer strife. Another factor influencing the pattern of world shipping is the availability and price of fuel at ports of call.

> A nuclear merchant ship can sail 2 to 5 years without taking on fresh fuel. This will give nuclear ships the limitless range of the sailing ship and the reliability and independence of the steamship. This freedom of operation could open new trade routes throughout the world.[79]

The finally and-- potentially -- the most significant factor is that nuclear powered engines do not require oxygen for their operations, thus making atomic fuel the perfect match for submarines. The use of commercial submarines would not only make for safer shipping, by freeing vessels from the tyranny of weather, but also open the arctic ocean to commercial sea transportation, thus considerably shortening the East-West routes.

---

[77] Ibid., p. 23. While conventional ships are also capable of such performance, their fuel consumption increases very rapidly as speeds increase, thus requiring "such large fuel tanks that the capacity for cargo will drop below the point of profitable return. Nuclear ships, on the other hand, require little fuel storage space." Ibid.

[78] Ibid., p. 4.   [79] Ibid., pp. 3-4.

207.

D. Aircrafts and Rockets

Having discussed transportation on earth and water let us briefly turn to air and space. Air propulsion is achieved either with the internal combustion or the jet engines, and fueled, respectively, by aviation gasoline and jet fuel.-- a "blend of lower grade gasoline, kerosine and distillate fuel."[80] Extrapolations from present technology seem to indicate that for the foreseeable future aviation will remain the captive market of oil.

> For a number of years the Air Force and the Atomic Energy Commission jointly sponsored a program aimed at developing reactors for the propulsion . of manned military aircraft. . . Early in 1961, after an expenditure of roughly $1 billion, the program was stopped on the grounds that 'the possibility of achieving a military useful aircraft in the foreseeable future is still very remote.'[81]

Having heard the verdict for military aircraft, one need hardly pause to consider commercial aviation. With a few possible exceptions[82] indications are that air transport will continue to be fueled by liquid hydrocarbons, even as aviation velocities reach the limits set by our planet's gravity. Past these velocities we enter into space, the uncontestable domain of nuclear energy. As present space travel grows from

---

[80]Schurr, Netschert, et al., op. cit., p. 118.

[81]USAEC, Nuclear Reactors, op. cit., p. 28.

[82]Possible exceptions may include giant nuclear-powered military aircraft, such as the C-5A, used as airborne command posts, and possible long-range commercial sub-orbital flights using rocket fuel, but their impact on total aviation oil demand is bound to be minimal

its embryonic phase, nuclear energy will be the midwife to its birth.[83]

In view of the crucial part played by mineral fuels in the past in making possible significant changes in transportation -- coal in the rise of the railroads, oil the the development of automotive transport -- it may be that at some date in the distant future nuclear fuels will be looked back upon as the energy source without which the revolutionary transportation system of the space age would have been impossible.[84]

IV. SUMMARY

The stated aim of this dissertation is to examine the relationship between oil and international relations during the time of their significant interaction. The aim of this chapter is the setting of the

---

[83]It has been seen that the richer the fuel in fissionable atoms the more compact the reactor. The paramount importance of weight in space travel forces the use of almost pure fuel (at $5,000 per pound of U-235). Nuclear reactors in space are presently used for satellite auxilliary power (SNAP-10A). Further improvements are expected for orbital labs, other satellites, and lunar bases. We then come to nuclear rocket propulsion (such as project NERVA) and nuclear electric propulsion (SNAP-50) as programed for the Mars manned missions. Improvements in these two applications should be sufficient for interplanetary travel. For interstellar travel (Alpha Centauri, our nearest star, is 4.3 light years away) we will enter the field of today's science fiction which may probably be tomorrow's reality. This will include such items as matter-antimatter reactors, gravitational control, etc., which will test the universal validity of Einstein's theory. See USAEC, Nuclear Reactors, op. cit., pp. 28-33; Direct Conversion of Energy, op. cit., p. 32; and SNAP Nuclear Reactors (see p. 159)(Oak Ridge, Tenn.: USAEC, 1966), pp. 6-10.

[84]Schurr, Netschert, et al., op. cit., p. 28.

temporal framework of this interaction using the criterion of political significance. The underlying assumption is energy competition without undue government interference or, more precisely, without drastic changes in present government energy policy. It has been assumed that while oil demand would grow in absolute terms until the end of the century, that further oil demand in the following decades would depend significantly on technological developments in energy production and consumption. Throughout the analysis major emphasis was put on the limitations within which future technological developments could operate in the various energy producing and consuming areas. Let us highlight the major points of the discussion.

In regards to energy production technology it was stated that organic energy sources share the common trait of mining and bulk transportation, that these traits are a basic limitation to radical technological developments, and that future developments in production and transportation will presumably favor natural and synthetic oil. Passing to inorganic energy sources, we eliminated most contenders for various reasons: hydropower for lack of abundant reserves and due to alternative uses for water, geothermal and solar energy for lack of knowledge, and tidal and windpower for lack of significance. This leaves us with nuclear energy which poses greater challenges but fewer limitations to technological developments. Oil was demoted from the production of electricity, which will be increasingly produced from nuclear energy and, therefore, at progressively decreasing costs.

In the discussion of energy consumption technology, oil was subjected to the joint barrage of electricity and nuclear energy. In the residential sector oil was ousted by a combination of lower electricity prices and higher standards of living. In the commercial, industrial, and multiple dwelling sector, oil was set to compete with nuclear energy for the production of low-temperature, low-pressure steam. It was seen that in this sector the impact of future fission reactor technology will be limited by reasons of cost and safety.

In regards to the vital transportation sector it was noted that the displacement of oil from the motor vehicle market is contingent on the development of a competitive fuel cell or battery operated electric car. In the case of railroads, the diesel engine locomotive is either replaced by the nuclear powered locomotive or not installed in the future high-speed mass transit railroad systems. Oil loses its preponderance at sea but maintains it in the air, up to the limits of the stratosphere.

Having recapitulated the discussion, let us now concoct our own scenario. Let us reasonably assume the perfection of a competitive breeder reactor and of the magneto-hydro-dynamic generator in the 1980's. Let us also accept the estimate of the doubling of electricity demand every decade and of the equal sharing of the electric utilities market between nuclear and conventional thermal plants by the end of the century. Let us assume the increasing competitiveness of nuclear thermal stations, past a point where the only limitation to the construction of new nuclear thermal stations for most base load demand is the constraint set by fissionable material reserves. Let us further assume the competitiveness of fusion reactors with breeder reactors by the year 2000. From then on, and assuming a vast, efficient and dense grid system, the construction of new conventional thermal stations will be limited to the satisfaction of daily peak load requirements, while the existing ones will be phased out after the end of their useful life, say 20 years. The main technological competition for new base load electrical capacity will then be limited to the half-century old conventional nuclear fission and to the burgeoning nuclear fusion thermal plants -- a fit omen for the new millenium.

We now turn to the various energy consuming markets. Having been chased away from electric utilities, oil will suffer the same fate in the residential sector. Assuming the widespread use of air conditioning equipment, the gradual change over to electric heating will be accomplished largely by the elimination of seasonal electricity demand fluctuations, without a significant enlargement of base load capacity. Barring dramatic reductions in electricity prices oil will probably maintain its position in the production of low-temperature, low-pressure steam in the commercial, multiple dwelling and urban industrial sectors, but will come under increasing competition from natural and synthetic gas. But after the introduction of fusion reactors into urban centers oil will lose most of its markets in these various urban sectors as well.

In regards to the internal combustion engine, let us assume the perfection of the electric car in the 1980's, its competitiveness in price, performance and convenience in the 1990's, government legislation banning further construction of air-polluting motor vehicles by the year 2000, and the phase-out of the internal combustion engine in motor vehicles a decade later. This development will produce a large decrease in oil demand, but in view of the low conversion efficiency of the internal combustion engine, it will represent a correspondingly smaller increase in electricity demand.

Still in the transportation market, let us assume the gradual construction of electric-powered high-speed mass-transit systems between proximate urban centers, the perfection by the year 2000 of a competitive and safe nuclear-powered locomotive, and the corresponding replacement of the diesel locomotive in an additional decade. For vessels, let us assume the competitiveness of nuclear energy in large bulk carriers in the 1980's, in most passenger and passenger/cargo ships in the 1990's, and the phase-out of the conventional powered ships two decades later. With space as the monopoly of nuclear energy oil will then maintain its domination in the aviation market.

Assuming the general validity of the stated premises, we can then plot the following oil demand curve for the industrial countries. In view of growing energy demand in the various sectors, oil demand will grow in absolute terms up to the end of the century, remain stationary for the next decade, after which it will very sharply decline, as the full impact of previous technological developments in energy production and consumption come into effect. The decline will continue until the point where oil demand will again grow with the demand of its severely reduced markets. These would include rural areas unconnected to power grids, aviation, a small fraction of the various energy-consuming sectors, and the non-energy sector, mainly lubrication, the production of protein via bacteria, and the rapidly growing petrochemical industries.

Assuming that the above situation will be mirrored in developing country urban areas, and neglecting their low-energy consuming rural areas, we can then visualize a situation in about half a century where oil demand will be sharply curtailed, thus reducing its economic significance in international trade. Perhaps more important, the loss of oil's major functions will also deprive it of its present strategic significance and of its role as a political weapon, thus reducing oil to the rank of a "normal" primary commodity.

APPENDIX TO CHAPTER III

Figures A-1, A-2, and A-5, as well as the following text are taken from Glenn T. Seaborg, Fission and Fusion - Developments and Prospects; Figure A-3 is taken from Bernard J. Eastlund and William C. Gough, The Fusion Torch: Closing the Cycle from Use to Reuse.

Figure A-1

Figure A-1 illustrates the major components of one type of fusion power plant and the system's operation. This is a system based on a deuterium-tritium fuel cycle. In the case of this D-T cycle plant, neutrons from the fusion reaction are absorbed in a lithium "blanket" and heat the blanket, maintaining it in a molten condition. Thus the lithium acts as a heat transfer medium, a means of generating tritium for further use in the fuel cycle, and as a source of additional thermal energy. The molten lithium is then passed through a heat exchanger where it gives up its energy, along with some tritium, to potassium vapor. The potassium vapor drives a turbo-generator and then exits into a combined tritium recovery system and heat exchanger. This heat exchanger transfers much of the remaining energy in the potassium to water and generates steam for a steam turbine. The overall system thermal efficiency has been estimated

at 60 percent. Of course, part of the electrical power produced is
needed to run vacuum pumps and liquid metal circulation pumps, as well
as to provide the intense magnetic fields for the fusion reactor.

Figure A-2

Both the D-D and D-^{3}He cycles may be used for another type of
fusion reactor we are considering with great interest. That reactor
would employ the principle of direct conversion. Figure A-2 illustrates
the direct conversion of fusion power to electricity through what is
called the "mirror machine." This system has been proposed and is being
developed by Dr. Richard Post at the Lawrence Radiation Laboratory in
Livermore. Typically, the D-^{3}He reaction would be used in such a conversion
system. The principal energetic products are charged helium and
hydrogen ions.

The reaction products from the fusion reactor would escape from
the end of the magnetic mirror and their already low ion density would
be further reduced by expansion into a larger chamber. In doing so,
the rather random motion of the charged particles would be converted into
a directional stream. This expansion process is similar to that occurring
in the expansion of hot gases in a rocket nozzle. Electrons escaping
with the ions would then be separated electromagnetically from the positive
ions and the electrons would flow to "ground" of the electrical system.
The positive ions stream, carrying most of the energy of the fusion
reaction and consisting of particles of differing energies (or voltages)

would be caught by a series of eletrostatic collectors. Each collector would be kept at a different potential, with the first ones being at voltages less than the average of the particles and later ones at voltages equal to or higher than the average. Suitable voltage transformers would convert all of these voltages to a common potential, from which useful power could be obtained in the form of high voltage DC. I should point out that most of the particles would be at energies near the average potential. Only about 30% of the energy would have to be subjected to voltage transformation.

You will note that no intermediate conversion to heat is required in this type of system; the kinetic energy of the positive ions is converted directly into electricity. Since there are virtually no thermodynamic limitations, the efficiency of this process could be very high, of the order of 90%. The equipment would appear to be very simple to construct. Very low capital costs for the direct conversion equipment possibly as low as $10-15/KWe, appear to be obtainable for 100 MWe units. Please understand that this direct conversion concept is as yet untested. Although preliminary small scale tests of the concept will begin in fiscal year 1971, we have no plans as yet to build full scale units. In other words, direct conversion from fusion is not just around the corner. However, it could become an important element in future fusion power systems.

Figure A-4

Figure A-4 shows some of the things that might be accomplished using the Fusion Torch to process material. With this system, the plasma energy is used to vaporize and ionize solid material -- converting initially complex chemical compositions into an ionized gas consisting only of elements. The elements could then be separated by a variety of techniques. Toxic chemicals could be reduced to their basic constituents, ores reduced and alloys separated.

Figure A-5

Figure A-5 indicates some of the ways that the ultra-violet radiation from the Fusion Torch may be applied. Here again the potential is enormous. Among other things it includes the large-scale desalting of seawater, bulk heating for many applications, the sterilization of sewage and other wastes, food production through algae culture and possibly through the direct synthesis of carbohydrates from carbon dioxide and water. Through the production and use of ozone, the Fusion Torch has been suggested as a method of sterilizing drinking water, of reviving "dead" lakes and rivers by reducing their excessive organic matter and of reducing industrial air pollution.

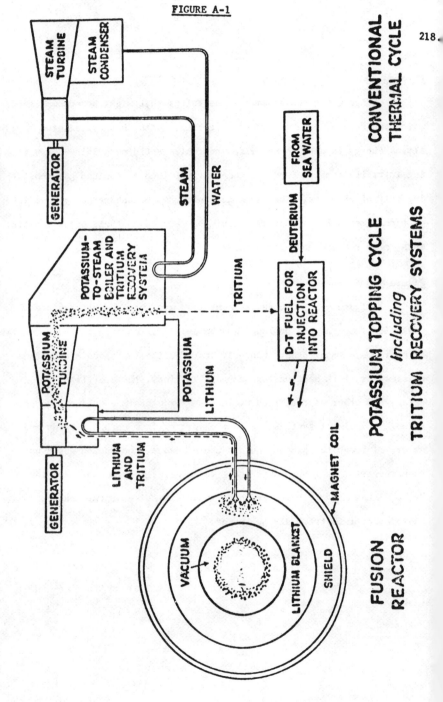

FIGURE A-1

FIGURE A-2

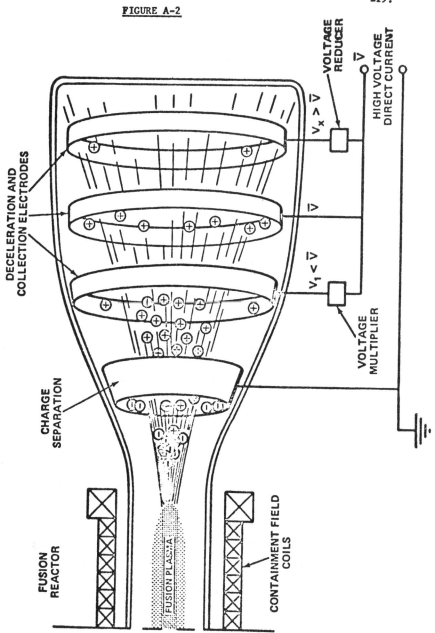

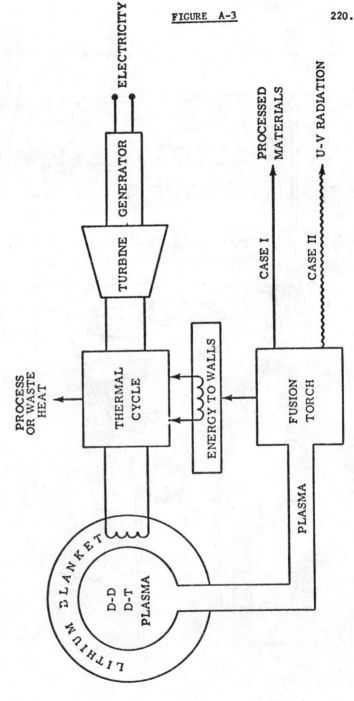

FIGURE A-3

220.

FIGURE A-4

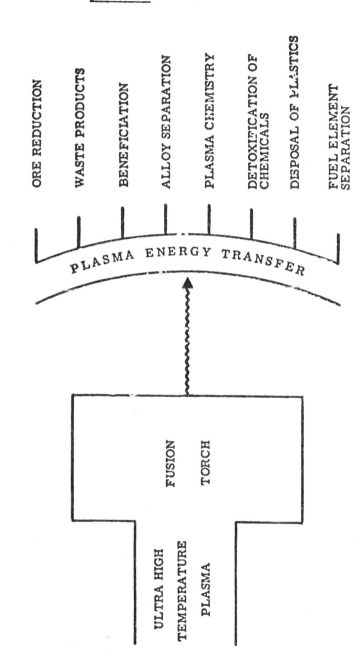

222.

FIGURE A-5

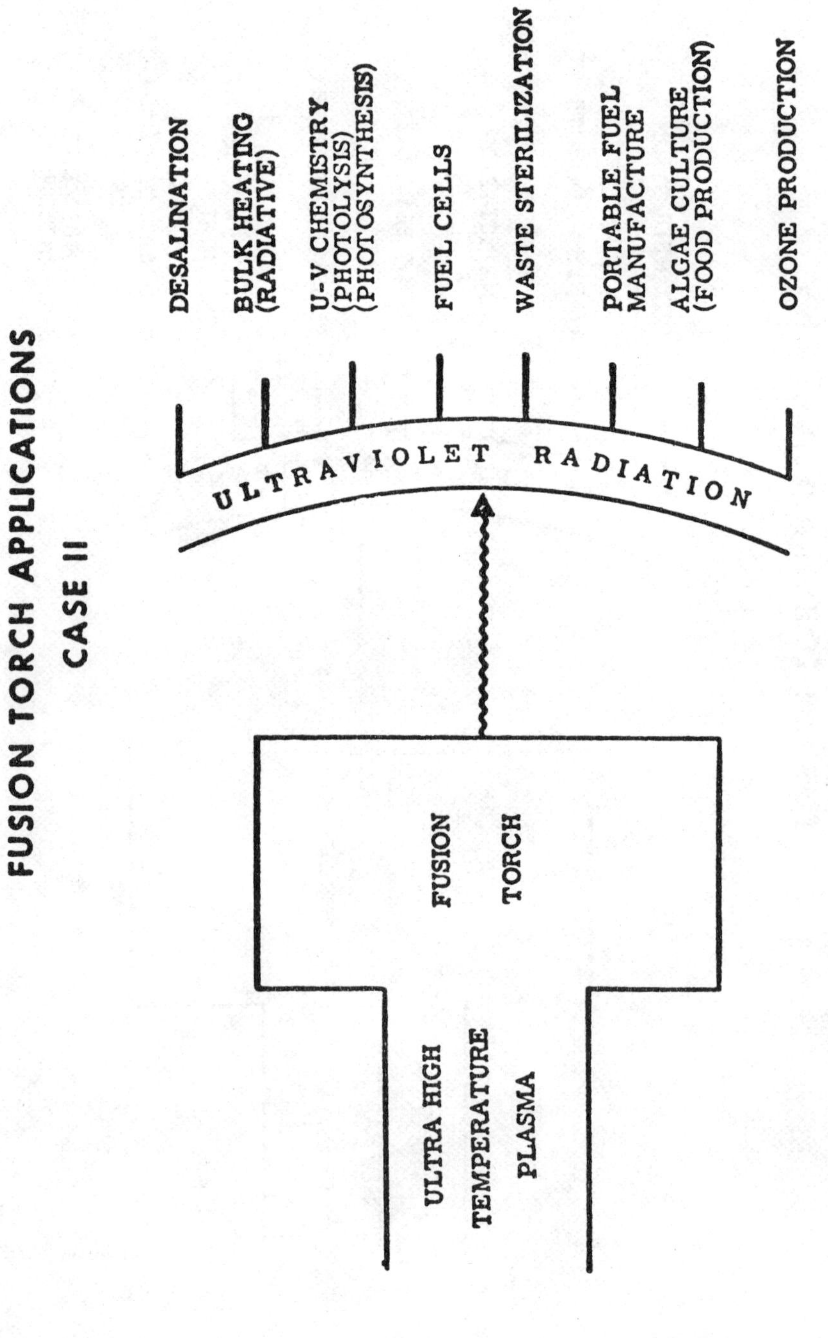

FUSION TORCH APPLICATIONS
CASE II

ULTRAVIOLET RADIATION

- DESALINATION
- BULK HEATING (RADIATIVE)
- U-V CHEMISTRY (PHOTOLYSIS) (PHOTOSYNTHESIS)
- FUEL CELLS
- WASTE STERILIZATION
- PORTABLE FUEL MANUFACTURE
- ALGAE CULTURE (FOOD PRODUCTION)
- OZONE PRODUCTION

FUSION TORCH

ULTRA HIGH TEMPERATURE PLASMA

PART II

OIL AS POLITICS

INTRODUCTION

Part I has delineated the main characteristics of oil in international trade. It is the task of Part II to integrate international oil trade with international relations. This then requires a framework that incorporates the desired variables within the chosen time period at a relevant level of abstraction. The variables are technology, and national and international economics and politics. The chosen time period is from the beginning of the oil industry until its depolitization. The level of abstraction should be one that best brings into relief the relations among the chosen variables. Let us rephrase the problem. How does international oil trade with its technological base fit into the maze of national and international economics and politics? For some writers the answer is simplicity itself. According to Fontaine:

> One may rest assured without fear of error that all the events of the past decades, including those that everyone today with confusion feels are weighting on the world, can very easily be explained by the battles - secret or public - for the possession of petroleum sources spread around the terrestial globe.[1]

Heady stuff, no doubt, which couched in proper academese could well compete with many theories of international relations; but we will

---

[1] Pierre Fontaine, *La Guerre Oculte du Pétrole* (Paris:Dervy, 1949), p. 9. My translation.

resist the temptation to pursue this line of thought and look for guidance elsewhere. If we must with reluctance reject the incorporation of international economic and political relations as a variable of the "battle for oil" we will have to try the reverse. Let us, therefore, examine how theory relates international trade to international politics. This could then give us a framework to analyze the role of oil in international relations. In order to set the parameters we will examine one theory from the economic and political interpretations of international relations.

According to Aron there are four ideal types of economic interpretations of international relations: mercantilism, liberalism, national economy and socialism.

> These schools define themselves and oppose each other by their interpretations of commerce (or of exchange) considered as the essence of economic life. According to the mercantilists, commerce is war, according to the liberals commerce is peace, on the sole condition that it be free. According to the national economists, commerce will be peace when all nations are developed; according to the Marxists commerce is war under capitalism, commerce will be peace with socialists.[2]

For the time being we will refrain from commenting on the relation, if any, between commerce and war, and examine the Marxist theory of imperialism. This theory is chosen because it incorporates most of our desired variables, as it attempts to explain the behavior of the various actors on the national and international scene, because it is the alleged vade mecum

---

[2]Raymond Aron, Peace and War. A Theory of International Relations (Garden City, N.Y.: Doubleday & Co., Inc., 1966), p. 253. Italics in the original.

of Communist states in their relations with the outside world, and, last and least, because of its large acceptance, albeit in mongrelized form, in the body of the "political" literature on the international oil industry.

According to Hoffmann general theories of international relations can be classified in the broad categories of theory as a set of answers and theory as a set of questions.[3] We have our own questions, and will therefore concentrate on the former. These again are subdivided according to the "realist" theory of international politics on one hand, and philosophies of history on the other. The latter being far too vague, we will choose the former.

The choice of Marxism and political realism as the parameters of our proposed framework is somewhat of a subterfuge, as parameters -- per definition -- imply extremes. Nevertheless, these theories will serve as excellent springboards for Part II. As will be seen, both theories share some common traits, but one emphasizes economics and society, the other politics and state. For one, commerce is paramount; for the other it is irrelevant.

I. THE MARXIST THEORY OF IMPERIALISM

In our day and age imperialism is the most popular term from the lexicon of international relations. And if not the most popular, then a

---

[3]Stanley Hoffmann (ed.), Contemporary Theory in International Relations (Englewood Cliffs, N.J.: Prentice-Hall, Inc., 1960), p. 29. Hoffmann makes a more thorough classification in The State of War, Essays on the Theory and Practice of International Relations (New York: Praeger, 1966), p. 16, but this will do for our limited purpose.

close second to the term colonialism. Perhaps the best way to start describing the present Marxist theory of imperialism is by differentiating it from its bourgeois homonym.

"Bourgeois ideologues" do not agree among themselves. To give an example, Aron defines it as "the diplomatic-strategic behavior of a political unit which constructs an empire, that is, subjects foreign populations to its rule,"[4] while Morgenthau sees it as "a policy that aims at the overthrow of the status quo, at a reversal of the power relations between two or more nations."[5] If the difference between these definitions is already quite pronounced the matter becomes more complex when one considers rule and power in terms other than military occupation. Indirect rule and power being notoriously fickle and subject to opinionated arguments, "Everybody is an imperialist to someone who happens to take exception to his foreign policies."[6]

Add to this the variable of domestic groups and nationalities who may feel oppressed by the domestic majority. "To trace the limits of imperialism clearly, the frontiers of nations would have to be visible on the map of cultures, of languages or of popular aspirations."[7] We can then rephrase the quotation to read "everybody is an imperialist to

---

[4] Aron, op. cit., p. 259.

[5] Hans J. Morgenthau, Politics Among Nations: The Struggle for Power and Peace, Fourth Edition (New York: Alfred A. Knopf, Inc., 1967) p. 42.

[6] Ibid., p. 41.

[7] Aron, op. cit., p. 260.

someone who happens to take exception to his foreign <u>and</u> <u>domestic</u> policies."[8]

"To add to the confusion, certain economic and political systems and economic groups, such as bankers and industrialists, are indiscriminately identified with imperialistic foreign policies."[9] We can, therefore, revise the popular bourgeois definition of imperialism to that of a household word where "everybody is an imperialist to someone who happens to take exception to what he is doing." This definition is an accurate reflection of the operative framework of some writers on the international oil industry.[10]

"Theories are always partially explained by historical circumstances."[11] The Marxist theory of imperialism came to the fore in the first two decades of this century as neo-Marxists tried to explain why Marx's predictions on the inevitable downfall of capitalism had failed to come about. Some revisionists, notably Bernstein, saw the reason in the evolution of capitalism. Others, notably Hobson, Bauer, Hilferding,

---

[8]This can lead to interesting chain reactions. Thus, U.S. policy in general and oil policy in particular may seem imperialistic to Venezuela, whose claims on more than half of Guyana's territory may smack of the same to the latter, which, in turn, has been accused of imperialism by some Amerindians. (I quote this accusation from memory, from a Venezuelan newspaper during the brief Amerindian revolt in January 1969).

[9]Morgenthau, <u>op. cit.</u>, p. 41.

[10]For another chain reaction, this time from the realm of oil imperialism, involving the U.S., Brazil and Bolivia, see: Lourival Coutinho, <u>O Petróleo do Brasil: Traiçaõ e Vitória</u>. Rio de Janeiro: Coelho Branco

Adler, Luxemburg, Lenin and Bukharin, noted the concentration of capital in industry and banking, the export of capital to colonies and Latin America, the scramble for Africa, later World War I, and tied these disparate elements together to form the classical theory of imperialism.

The classical theory is well known and hardly needs any elaboration. For our purposes we are more interested in its update, in the way that it explains the relations among industrial capitalist, communist, and developing nations. While the narrative is not central to the subsequent analysis it will be referred to in Chapter V. The script generally runs as follows.[12]

World War I was caused by the contradictions inherent in capitalism at its highest stage of monopoly. Its effect was to snap the weakest link from the chain of imperialism and the creation of the first Socialist state. The world depression was a normal phenomenon already analyzed by Marx. The ensuing pauperization of the masses generated widespread discontent against the capitalist system which was countered by fascism, by a further sharing with the masses of a part of the colonial profits, and by a spiralling of the armaments race, all of which paved the way for World War II.

The outcome of World War II resulted in the spread of socialism to Eastern Europe and to parts of Asia. The capitalists were faced with

---

1957; and Amado O. Canelas, *Petróleo: Imperialismo y Nacionalismo; Roboré Derrota de Dos Pueblos*. La Paz, Altiplano, 1963.

[11]Aron, op. cit., p. 253.

[12]What follows is a somewhat ironic resume of the vulgate Soviet literature on the subject.

the impoverishment and the discontent of their masses and were forced to implement to so-called welfare state with its concomitant state capitalism in many branches of industry and in those sectors directly affecting the public interest. This loss of profits required further exploitation from the colonies, but the latter -- encouraged by the successes of the socialist countries -- demanded independence.

The European ruling classes were finally forced to grant the independence to their colonies under the combined pressure of the United States, which wanted to increase its share of the colonial market, some colonial wars of liberation, and the discontent from their masses who wanted to avoid expensive colonial wars. But the granting of political independence was only a retreat into a new form of exploitation -- neo-colonialism -- which combines the worst features of colonial exploitation with a lack of political responsibility.

The present day intricate transmission belt runs as follows. The imperialist governments appropriate money from their masses via taxation in order to provide for foreign aid and loans, and to pay for infiltration apparatuses such as USIS, CIA, the Peace Corps, etc. Foreign aid is used to influence, if not to bribe, the foreign leaders, while the repayment of high interest loans must be squeezed out from the local masses. The resulting discontent is abated by massive propaganda, snuffed in the bud by foul means against individuals and/or groups or, failing that, by military intervention. This political manipulation is then used for direct economic exploitation.

The capitalists pay low wages to the native workers, and low prices for exported raw materials. In return, they charge high prices for manufactured products and services, such as transportation and insurance. A part of these super-profits is then passed on to the masses of the imperialist nations in the form of higher wages and welfare state services. Another part goes for the armament race and for the payment of mercenaries. Thus, the imperialists manage to make high profits and to lull their own masses, thereby postponing the day of the inevitable revolution.

The remedy out of this vicious circle is to cut the gordion knots: oust the monopolies, cease all truck with Moloch, and proceed towards the non-capitalist road of development with assist from the socialist countries. This will not only benefit the masses of the developing nations but also accelerate the downfall of capitalism in the industrial nations, thus ushering in the era of world peace.

We will not engage in a rebuttal of this nursery tale as the challenge of tautologies is a rather thankless task. In any event it would be outside the scope of this thesis. What we are interested in is the underlying assumptions and the linkage between the relevant variables. The basic assumption is that the substructure determines the superstructure: the forces of production determine the relations of production, which are always class relations, which in turn determine a nation's domestic and foreign policies. "The whole maze of international policies seems to be cleared up by a single powerful stroke of analysis."[13] In fact the blurring of the lines between national and international economics and politics degenerates into a tautological exercise of qui bono across the whole spectrum which can only be accomplished by delving into motives and with a heavy assist from conspiratorial explanations.

Let us go down along the line, starting with technology. According to the labor theory of value, labor is the only source of profit. Technology results in labor-saving devices and in the accumulation of capital. "Accumulation accompanied by qualitative change in the composition of capital is . . . a remedy which though alleviating for the moment the situation of the individual capitalist makes matters worse in the end."[14] In the meantime, technology is exploitation in capitalist countries and rationalization in socialist ones.

From the substructure we turn to the superstructure, to the relation between national economics and politics. Needless to say, the politics of socialist countries bask, per definition (no capitalists, no contradictions) in a state of harmony of interests. It is another story for the "committee for managing the common affairs of the bourgeoisie." First the means. The electoral process, what Lenin referred to as the politics by arithmetic, is -- at best -- a sham. Economic power determines political power. The monopolists hold the key levers of influence, including the

---

[13] Joseph A. Schumpeter, Capitalism, Socialism, and Democracy, Third Edition, Harper Torchbooks (New York: Harper & Row, 1962), p. 51.

[14] Ibid., p. 49.

mass media to detract everyone's attention from his true class interests, and money to bribe the politicians and their parties. Now the ends: everything that the governments do; for their actions are performed either in the direct interest of the ruling classes or out of fear of discontent from their masses.

> Again, in practically all cases the theory can be made tautologically true. For there is no policy short of exterminating the bourgeoisie that could not be held to serve some economic or extra-economic, short-run or long-run bourgeois interest, at least in the sense that it wards off still worse things. This, however, does not make the theory any more valuable.[15]

Matters tend to be more complicated in the relations between sub- and superstructure in developing countries. In short this deals with the Communist theories of development. The cruder forms are reflected in the previously described scenario. The more scholarly attempts back away from single-cause analysis into something resembling multi-cause Western efforts, albeit with rationalizations of the scriptures by the use of qualifiers, such as "objective" and "subjective" behavior, etc. We can leave Soviet Academicians with the task of fitting tribalism into the Marxist framework and turn our attention to the relations between national and international economics.

---

[15] Ibid., p. 56.

Let us start with the foreign trade among socialist countries. These are, again by definition, characterized as beneficial to all parties concerned and, never mind about Rumania, Albania, or China, as a socialist division of labor. By the same token, the absence of Soviet contradictions graces the relations with its trade partners, be they capitalist developed or developing nations. But this is not necessarily true of the foreign trade of Eastern Europe with capitalist developed nations, as this might endanger the Socialist Commonwealth (e.g., Czechoslovakia, West Germany), or with developing nations in commodities that compete with Soviet export. Until very recently, the prime example of this was oil. <u>Quod licet jovi non licet bovi</u>.

Trade among developed capitalist nations is, so to speak, a traffic in contradictions. According to the former redeemer Kwame Nkrumah, ". . . perhaps the most ludicrous is the constant traffic in the same kinds of goods, products and commodities between countries. Everyone is busy, as it were, taking in the other's washings."[16] But this is not representative of socialist economic analysis which, with one eye on Pareto and the other on Lenin, grants the short-term benefits to the respective economies and consumers but emphasizes the dire consequences of the long-term outcome of the battle between the monopolies.

If foreign trade with socialist countries is a harmony of interests, and among developed capitalist countries a traffic in contradictions, then

---

[16] Kwame Nkrumah, <u>Neo-Colonialism, the Last Stage of Imperialism</u> (New York: International Publishers, 1966), p. 184.

235.

trade between developed capitalist and developing nations is exploitation.

> At (the) August 1962 conference of Marxist economists . . . the principal paper estimated that the developed nations were getting $14 to $16 billion annually from the underdeveloped by means of "unequal exchanges."* Since the total exports of the developed countries to the underdeveloped amounted to around $20 billion in 1961,** this implies either that their "equitable" value was only $4-$6 billion, a 400 per cent overcharge, or else that the "equitable" value of the underdeveloped nations' exports was really around $34-$36 billion, a markdown of 40 per cent. It is a round sum either way: "Si vous faites cela vous ne ferez pas peu."[17]

---

*Current Digest of the Soviet Press, September 19, 1962, pp. 13-14.

**OECD, Foreign Trade, Series A, August 1962. The "developed" nations are taken as the United States, Canada, Japan, OECD Europe, Australasia, South Africa, and Israel; the Soviet and the Chinese blocs are excluded; the rest of the world is taken as "underdeveloped."

Finally, we come to the capstone of the theory, the relation between national economics and international politics, to the causal effect between capitalism and war.

> This is, in short, single-cause explanation in one of its less impressive forms. . . . Capitalism equals war because it allows minority interest to determine majority will; socialism equals peace because it is rule in the interest of the people at large. Under socialism, the archetypes of capitalism villainy will be done away with - special interests, if they exist

---

[17]Adelman, The World Oil Outlook, op. cit., p. 111.

at all, will no longer be able to corrupt the
rational processes of the state.[18]

Comments are superfluous: one does not argue with a credo, but one may point to a flaw of logic.

To say that capitalist states cause war may, in some sense be true; but the causal analysis cannot simply be reversed, as it is in the assertion that socialist states mean peace, without first making sure that the causal analysis is complete. Is it capitalism, or states, or both that must be abolished?[19]

The above reasoning with minor adaptations should also take care of all other economic interpretations of international relations. At this point one could well ask what, if any, is the role of oil in these proceedings. Specifically, what is the Marxist interpretation of <u>oil imperialism</u>? In his discussion of various authors of the 1920s -- when this term first came into use -- Foursenko states:

All the above writings concentrated their attention on the role that the battle for oil acquired in world politics. But as a rule these works contained errors and obvious exaggerations. They declared oil as the main factor in world politics and even asserted that the beginning of the next world war depended on the development of the rivalries for oil. . . At that time the famous Soviet historian F.A. Rotstein warned against the

---

[18]Kenneth N. Waltz, <u>Man, the State and War: A Theoretical Analysis</u> (New York: Columbia University Press, Paperback Edition, 1965), pp. 146-147. Italics in the original.

[19]Ibid., p. 157. Should the uninformed believe that the Sino-Soviet conflict requires a reformulation or at least a qualification of this tenet, let him be promptly disabused. Both sides ascribe the root of each other's sins to economic causes that deflect them from the inherent harmony of interests among socialist states. The Soviets accuse the

attempts to "reduce the rivalries between the imperialistic powers to one main center - oil." "In spite of the huge importance of oil . . . neither coal nor steel have lost any of their importance and may just as easily as oil be the object and cause of world conflict."* It is, therefore, quite evident that the term "oil imperialism" must be taken with ample reservations.[20]

---

*See the introduction by F. Rotstein in the book: E. Schultze, The Battle for Persian and Mesopotamian Oil (Moscow, 1924), p. 3.

Having thus restored our sense of perspective, what can one conclude on the Marxist theory of imperialism? Mainly that if one happens to be neither a Marxist nor a Soviet propagandist one will have to look somewhere else for guidance. One should, however, keep in mind that if to an identical hostility one adds a quest for dark motives and a penchant for conspiracies, and if one substitutes "free enterprise" for "class struggle," "democracy" for "socialism," and "common Western heritage" for "proletarian internationalism," one can arrive at the same analythical method and at identical conclusions, albeit with opposite signs. But with one difference: to the best of my knowledge the John Birch Society does not claim possession of the key to the understanding of the human experience.

Chinese of destroying the Communist Party during the Cultural Revolution, thus releasing all kinds of evils to the surface, while the Chinese flatly assert that since Khrushchev the Soviet Union has "gone bourgeois," with all its foreign policy implications.

[20]A.A. Foursenko, Neftianie Tresti i Mirovaia Politika, 1880-godi -1918g. (Moscow: Nauka, 1965), p. 191. My translation.

II. POLITICAL REALISM

The roots of political realism are embedded in the time-honored concept of Realpolitik. This term is nearly as difficult to define as the term imperialism. According to Waltz:

> Realpolitik is a loosely defined method, which is described as being necessary when a given purpose is sought under a specified condition. The purpose is the security of states and the condition, anarchy among them.[21]

In deference to its teutonic roots we could thus refer to Realpolitik as the Primat der Aussenpolitik and seek its theoretical justification in Treitschke's Machtpolitik. And in reference to its late nineteenth century flavor we could also connect Realpolitik with the "national economy" theories of F. List, with German protectionism in general, and of the vital industries, the so-called Commanding Heights in particular. But both authors lose in translation: not only to English but also to the present international scene with its near bipolarity, nuclear weapons, and spate of sovereign developing countries. We will, therefore, have to look at more recent editions and, as a matter of convenience, to American authors.

"In crossing the Atlantic, in becoming power politics, Treitschke's Machtpolitik underwent a chiefly spiritual mutation. It became fact, not value."[22] And as an explanation and prescription of fact it took the mantle

---

[21] Waltz, op. cit., p. 216.

[22] Aron, op. cit., p. 592. Italics in the original.

239.

of political realism. Which presents some problems, as "everyone, of course, thinks his own theories are realistic."[23] Nevertheless, political realism is firmly identified in the American academia with the theory of Morgenthau, to which we now turn our attention. Rather than to summarize his theory, a long and for our purpose unnecessary task, we will restrict ourselves to a capsule analysis of his underlying assumption, epistemology and operative framework, compare them with Marxism, and test the relevance of the model of political realism to our topic.

Let us begin with the underlying assumption. In the Marxist analysis man's transformation of nature results in forces of production that determine the relations of production. The ensuing society substructure is the main determinant of the human experience. Political realism, on the other hand, starts on an implied basis of the "animus dominandi, the desire for power."[24] Accordingly, the political realist "thinks in terms of interest defined as power, as the economist thinks in terms of interest defined as wealth; the lawyer, of the conformity of action with legal rules; the moralist, of the conformity of action with moral principles."[25] According to Hoffmann:

> Now, the decision to equate politics and power would be acceptable only if power were analyzed, not as a limited and specific set of variables, but as a

---

[23]Waltz, op. cit., p. 20.

[24]Morgenthau, Scientific Man vs. Power Politics, as quoted and weighted by Waltz, op. cit., pp. 34-35.

[25]Morgenthau, Politics Among Nations, p. 11.

> complex and diffuse balance between the variables
> with which the social sciences are concerned.
> Political man should properly be seen as the "inte-
> grator" of moral man, economic man, religious man,
> and so on -- not as a creature reduced to one special
> facet of human nature. Unfortunately, such an
> Aristotelian position is not adopted (by Morgenthau)[26]

Thus, in contrast to historical materialism, that views man's actions primarily as a response to his economic environment and analyzes him accordingly, political realism sees man as a seeker of power and analyzes his actions through the lens of politics. This basic assumption reflects itself in the epistemology, for as a result of the above "The master key (to the theory) is the concept of interest defined in terms of power."[27]

> Even if the role of power were as determining as the
> theory postulates, the question arises whether any
> scheme can put so much methodological weight upon one
> concept, even a crucial one; for it seems to me that the
> concept of power collapses under the burden. It is
> impossible to subsume under one word variables as
> different as: power as a condition of policy and power
> as a criterion of policy; power as a potential and power
> in use; power as a sum of resources and a power as a set
> of processes. Power is a most complex product of other
> variables, which should be allowed to see the light of
> the theory instead of remaining in the shadow of power.[28]

Add to the above that "The word power, in English, has a very broad (or very vague) meaning, since depending on cases, it translates the three French words pouvoir, puissance, force."[29] Compound this

---

[26] Hoffman, op. cit., p. 31.

[27] Ibid., p. 30.

[28] Ibid., p. 32.

[29] Aron, op. cit., p. 595.

with Morgenthau's definition: "Power may comprise anything that establishes and maintains control of man over man."[30] and we find that by a somewhat different route, political realism approaches the most objectionable feature of Marxist theory; epistemology based on tautology. For insofar as a theory is based on a timeless and all-embracing definition of its crucial concept it finishes by explaining nothing, or -- if that seems too strong -- by explaining only similarities. "The price one has to pay for identifying the 'timeless features' of the political landscape is the sacrifice of understanding the processes of change in world affairs."[31]

As we will shortly see, apart from single-cause analysis historical materialism and political realism share an operative framework based on nineteenth century experience. For the former the starting point is the travail of the industrial revolution; for the latter it is the era of relative peace between Vienna and Versailles. This operative framework is reflected in the model for analysis.

> The model of the "realist" is a highly embellished ideal-type of eighteenth and nineteenth century international relations. This vision of the golden age is taken as a norm, both for empirical analysis and for evaluation.[32]

As in the case of most general, single-cause theories, the model of political realism ecompasses most, if not all, of the variables germane

---

[30]Morgenthau, op. cit., p. 9.

[31]Hoffmann, op. cit., p. 35.

[32]Ibid., p. 32.

to its realm. As such it is a most impressive model with multiple variables woven into a rich tapestry. But the weight that the theory gives to the variables mirrors the international relations of the nineteenth century international system. Insofar as the various international systems are discussed at some length in Chapter V we will only mention here that the nineteenth century was characterized by a relatively stable balance of power, transnational homogeneity, minority rule via absolutism and/or voting property qualifications, and the corresponding separation of state from society. As a result of the above, the major powers were playing ". . . for limited ends, with limited means, and without domestic kibitzers to disrupt the players' moves,"[33] cabinet diplomacy encouraged rationality, and *laissez-faire* reigned supreme. By transplanting the weight given to these variables in the nineteenth century to our present international system, political realism shows a nostalgia for the good old days, somewhat akin to the Marxist deploring the passing away of the international solidarity of the working class.

This aberration can be seen in the political realist's view of the role of oil in international relations. Not surprisingly, Morgenthau analyzes the present role of oil through the lenses of nineteenth century politics. He assesses technological change and the consequent growing importance of oil in industry and strategy, but emphasizes mainly the latter. He sees its main significance in the fact that "The emergence of

---

[33] *Ibid.*, p. 33.

of oil as an indispensable raw material has brought about a shift in the relative power of the leading nations."[34] The role of oil in international relations is thus limited to the struggle among the leading nations for control over oil deposits, which -- as geology would have it -- are mainly located in the Near East. In regards to these deposits Morgenthau states:

> Control over them is an important factor in the distribution of power, in the sense that whoever is able to add them to his other sources of raw materials adds that much strength to his own resources and deprives his competitors proportionately. It is for this reason that Great Britain, the United States and, for a time France have embarked in the Near East upon what has aptly been called "oil diplomacy"; that is, the establishment of spheres of influence giving them exclusive access to the oil deposits of the region.[35]

Consequently, and with minor qualifications, " . . . the importance of the Arab states derives exclusively from their control of, and access to, regions rich in oil."[36] So much for the role of oil in international relations.

It would be unfair to build up a case on the basis of a few paragraphs. Therefore, we will only mention that in the above quotation the neglect of economics distorts the analysis and leads to misleading conclusions. Thus, the assertion that the major Western Powers control the Middle East oil deposits in order to add to their resources may vaguely

---

[34]Morgenthau, op. cit., p. 111.
[35]Ibid., pp. 111-112.
[36]Ibid.

reflect government motivation during the inter-war period but is quite off the mark in today's situation. With U.S. oil import quotas, present world oil surplus, and the predominant role of the major oil companies as international middle men, such an assertion flies in the face of evidence.

By the same token, the present importance of Middle East oil lies not so much in its abundance but rather in its low cost and proximity to the main consuming markets. Thus, the Middle East would lose much of its importance if its oil production costs were substantially higher than of the Canadian Athabasca tar sands, or if the North Sea hydrocarbon deposits were oil instead of gas. In other words, Morgenthau's analysis is based on a period of scarcity of oil and the result from a scarcity of variables.

But this need not detain us. Let us accept Morgenthau's verdict on the limited role of oil in international realtions and, by indirection, of the importance of this thesis. What we are after is the relevance of the model of political realism for an understanding of our subject. We can begin with the phenomena that it does explain.

First, the theory's definition of imperialism explains some of the actors' goals and methods. Thus, the goal of local preponderance (spiked with hopes of economic gain) explains the envisaged tug-of-war between Iran, Iraq and Saudi Arabia for control over the Persian Gulf after Britain's stated withdrawal in 1971. In addition, the method of economic imperialism, genus "oil diplomacy," explains the actions of the Great Powers during the inter-war period and, by stretching the point, presents U.S. influence over Venezuela.

Second, the theory explains the isolated cases of the economic causes of war with oil as a stake among neighbors. Thus, "The Chaco War is considered by some to have been primarily a war between oil companies."[37] Granting the above boon to the Marxists, and leaving oil companies aside, this would also explain the 1961 Kuwait-Iraq dispute and Britain's role in it.

Third, the theory partially explains the present importance of major oil exporting countries in world affairs. Accordingly, it would be safe to presume that, ceteris paribus, the absence of oil reserves would relegate Libya to the status of Sudan, Iraq to that of Afghanistan, and Kuwait to that of Burundi. And this just about exhausts the range of explanations. For this limited view of oil as a source of industrial and strategic power can only explain that, there being oil "haves" and oil "have nots," and nations being the power seekers that they are, there results a situation of conflicting national interests that can lead to national conflicts. To revert to the above example the importance of Kuwait in regional affairs is not so much its control over the Burghan field but rather what it does with the money it receives from the oil. To draw the list of phenomena that this theory does not explain would

---

[37]Ibid., p. 46.

draw this introduction out of all proportions. We will therefore restrict ourselves to the minimum by indicating the lack of adequate relations among variables.

First, the divorce between domestic and foreign politics. As an example, the slogan "Arab oil vs. Jewish votes" deserves the respect due to all slogans, but its total neglect hampers the understanding of U.S. Middle East policy. By the same token the theory would draw a blank in its attempts to explain the constant policy reversals of Argentina in regards to oil concessions to foreign companies, and the resulting effects on its foreign policies.

Second, and as a consequence of the first, the neglect of the relations between domestic economics and politics. Thus, the U.S. oil import quotas cannot be fully understood without reference to the "independents" or to the Texan legislature; neither can the German taxes on fuel oil, with the resulting impact on German oil imports, without considering the political strength of the Ruhr coal miners.

Third, the virtual exclusion of international economics from the realm of international politics. The balance sheets of oil companies may concern only their shareholders, but their effects on balance of payments may possibly affect the vital interests of nations.

Fourth, the inadequate explanation of the relation between foreign policy proper and foreign economic policy, as "oil diplomacy" is not restricted to "economic imperialism"; thus Iranian foreign policy in general, and towards Israel and South Africa in particular, cannot be understood without reference to its oil exports.

Finally, and as a consequence of all the above, the vague determination of national interest. The theory may explain why successive Arab governments have failed to nationalize the oil companies in the face of domestic pressures, but it cannot explain why changes in regimes, such as in Iraq or recently in Libya, have altered the nature of the political relations with most Western Powers.

All in all, we may conclude that the realist theory does not provide the map to travel from the placid nineteenth century cabinet diplomacy to the bedeviling present oil diplomacy. Marxism, as opposed to, say, Soviet foreign policy, has never really digested nationalism except by the facile method of identifying the interests of most nations and classes with the interests of a few nations with Communist regimes. Political realism has failed to adopt the offspring of the marriage between state and society, and, one may add, to understand the nature of the energy that this youth produces and consumes.

III. ALTERNATIVE METHOD

This introduction has so far followed the all too familiar ploy of criticizing the extremes in order to justify the subsequent claims to balanced judgment. But to the ensuing <u>mea culpa</u> one should add that the discussion has not only proved, if proof were needed, the pitfalls of single-cause analysis, but has also shown the complexity of the subject and, above all, the dangers of generalizing from unique historical circumstances.

248.

Yet the temptation to follow such a course of action is great. For if we conceive of many variables with multiple causation whose relations and causation differ according to levels of abstraction and conceptualization, and in time, we are faced with the unpleasant alternative of plodding in a highly complex model with rigid assumptions -- à la systems analysis -- or of plucking from a mishmash, as in most of the "political" literature on oil.[38] Add to this the fact that most models tend to favor some interests as opposed to others and, for no other reason, óne is tempted to agree with our Marxist colleagues on the class content of the social sciences and to retreat to the haven of ideology and tautology.

But this should be no cause for undue worry. As the gallant French say, the most beautiful woman can only give what she has. So can the social scientist , certainly this one. Therefore, calling Max Weber and Aron to the rescue, we will agree with the former on the need to accentuate the factors and variables which are of the writer's interest, and with the latter that "We must determine the focus of interest, the proper significance of the phenomenon or of the action that constitutes the nucleus of this specific domain."[39]

In view of the fact that "Of all the commodities in international trade, oil is undoubtedly the supremely political one,"[40] and that in this realm the supreme political act is the oil boycott, it seems to me that

---

[38] For a prime specimen of the above, skim through: Paul Friedrich, Oel Gegen Kriegsangst. Olten: O. Walters, 1951.

[39] Aron, op. cit., p. 5.

[40] Michael Tanzer, The Political Economy of International Oil in

it can serve as the focus of interest and as the nucleus of the oil domain, because "It is only in those extreme situations, involving oil boycotts, that the enormous latent political and economic forces always involved in international oil surface fully.⁴¹

Needless to say, the oil boycott, whose definitions will have to wait for the next chapter, is a means towards an end. The focus on the means alone could lead to Gupta's conclusion that "oil is not an unmixed good. It has also its satanic side,"⁴² and perhaps to his recommendation that "the use of oil for aggressive ends, under the Charter of the United Nations, should be outlawed."⁴³ But seen as a means toward an end, and focusing on states as the prime actors in international relations, the oil boycott lies at the junction of general foreign policy and of foreign economic policy.

In this respect the model proposed by Cohen, albeit with some reservations and adaptations to oil, seems to be quite fruitful. Pending further discussion in the next chapter, he proposes an analytical framework in which ". . . the overall objective of foreign economic policy is the joint maximization of the state's current income and its net influence

---

the Underdeveloped Countries (Boston: Beacon Press, 1969), p. 319.

⁴¹Ibid.

⁴²Raj Narain Gupta, Oil in the Modern World (Allahabad: Kitab Mahal, 1949), p. 3.

⁴³Ibid., p. 147.

in international affairs."[44] Economic influence is based on the "power to withhold" -- the essence of the oil boycott -- while the state's current income can be manipulated by means of: (1) commercial policy, (2) foreign investment policy, (3) foreign aid policy, and (4) balance of payments policy. These policies are defined in the broadest sense to include practically everything that a state does to affect its foreign economic relations, and what these policies do not include we will redefine so that they do. This latitude will then allow us to include in these policies most of our desired variables. This is the scope of Chapter IV. The time span that it analyzes is the post World War II period.

The inclusion, however qualified, of technology, geology, oil economics, etc., and of national and international economics and politics under the heading of foreign policy, however defined, is useful for the purpose of analyzing the focus of international oil. But this legerdemain will clearly not do in the analysis of the role of oil in international relations. If we are to avoid most of the excesses previously deplored in single-cause analysis, the generalization from unique historical circumstances and, above all, the tendency to exaggerate the importance of one's topic, we will have to reverse the procedure. Accordingly, Chapter V attempts to incorporate Part I, specifically the structure of the international oil industry, and the relevant findings of Chapter IV, into the framework of the international system, from the beginning of the oil industry to its presumed depolitization.

---

[44]Benjamin J. Cohen, Editor, <u>American Foreign Economic Policy, Essays and Comments</u> (New York:Harper &Row, 1968), p. 20.

CHAPTER IV

OIL AND FOREIGN POLICY

I. INTRODUCTION

As stated in the introduction to Part II the framework of this chapter is based on the model proposed by Cohen in his essay "American Foreign Economic Policy: Some General Principles of Analysis." In the introduction to this essay, Cohen states:

> Most economists and political scientists act as if 'never the twain shall meet." In their surveys and studies of international affairs, neither economist nor political scientist have devoted much serious thought to developing a systematic conceptual framework of analysis that would permit discussions of the allocation of scarce resources in support of power relations. The purpose of this essay is to start to lay the foundation of such an analytical framework.[1]

Foreign policy, according to Cohen, results from a combination of particular and general interests. "Foreign policy is always a function of specific lesser interests within the nation, the offspring of the interplay of powerful institutions, each trying to achieve its own particular ambitions and goals."[2] These lesser interests are mainly responsible for

---

[1] Cohen, op. cit., p. 1.

[2] Ibid., p. 3.

the irrationality observed in foreign policy. But the specific interests are and must be subordinated to the national interests, "the most basic of which is self-preservation - survival."[3] The duty of each state is to translate this quest for survival into a foreign policy strategy. "But this is a difficult matter, for the concept of national security is not a precise, meaningful guide for action; it is subjective rather than objective in content and consequently rather ambiguous."[4] To add to the complexity national security also involves "the preservation of minimum 'core values,' notably cultural survival and "a certain range of previously acquired values, such as rank, prestige, material possessions, and special privileges."[5]

> Pushed to its logical conclusion, such extension of the range of values to include more and more marginal values does not stop short of the goal of complete world domination. . . . However, most governments do not push the logic of national security so far and rely instead on less ambitious strategies of foreign policies.[6]

The reason why most governments refrain from such ambitious strategies springs not from moderation but from the balance of power, as "ultimately, national power sets the limits to the nation-state's choice of a strategy of foreign policy. . ."[7] Consequently, the state's "rational

---

[3] Ibid.

[4] Ibid., p. 4.

[5] Ibid., p. 5.

[6] Ibid.

[7] Ibid., p. 8.

solution is to broaden its range of options -- that is to maximize its power position, since power sets the limits to the choice of strategy."[8] Power is both end and means and "represents the ability to control or at least influence the behavior of others."[9] The means determining the ends, "national power can be accumulated to the extent permitted by the resources of the state," while "the available resources must be qualitatively appropriate to the chosen ends."[10]

> In a real sense, therefore, national power not only sets the limits to the selection of proximate foreign-policy goals; it also provides the instruments for their achievement. This is the sense in which the two basic problems of foreign policy are interrelated; through the sum total of national resources that can be employed to influence the objective of national security. Nowhere is this interrelation more apparent than in the determination of that subset of general foreign policy labeled foreign economic policy . . . [which,]
> . . . represents the sum total of actions by the nation-state intended to affect the economic environment beyond the national jurisdiction.[11]

Foreign economic policy is a "hybrid". On the one hand it is concerned with the allocation of scarce resources. Its short-term goal is to maximize national income while its long-range goal is to maximize national wealth. "On the other hand . . . it . . . is also concerned

---

[8]Ibid.

[9]Ibid., p. 9.

[10]Ibid., pp. 9-10.

[11]Ibid., pp. 10-11.

with national security. Its short-range goal is to provide maximum support for the state's general foreign policy, while its long-range goal is to maximize national power."[12]

> Even though foreign economic policy is concerned with the maximization of national power, it can actually operate on only . . . national _economic power_ . . . (which) . . . represents the ability to control or influence the behavior of others in _economic_ matters. The possibility for influence in economic matters derives from the fact that the world economy . . . is in fact a system of interrelationships in which to a greater or lesser extent every nation is dependent on all the others - dependent for commodities and services of various kinds, for markets and investments, for technology and skills. These dependencies are tolerated because . . . these benefits enrich each nation-state and increase its material wealth. The price to be paid for these gains is dependence on others. _The dependence of one state on another gives the latter influence through its control over that for which the former depends on it._[13]

In other words, the varying dependence of states on the international specialization of labor results in dependence on other states, each bent on the maximization of national power and on the use of such economic dependence for its political ends. Consequently, wealthy states will try to increase the dependence of other states on them in order to maximize the "power to withhold", while poorer states will attempt to resist such efforts.

---

[12] _Ibid._, p. 11.

[13] _Ibid._ Italics on _economic_ in the original. The other emphasis is mine.

But " . . . wealth per se is not sufficient to
exercise effective power in international relations;
national wealth and national power are not in fact
identical. . . . What matters is how that wealth
fits into the overall distribution of dependence and
influence in international affairs. . . . Over a
broad range of policies, there can be no doubt that
the two objectives of national wealth and national
power are functionally equal, complementary rather
than competing. . . . But there is also no doubt
that over a certain range of foreign economic policies,
the two objectives are in direct conflict. Within this
range a choice must be made ...."[14]

As Cohen concludes: "This in reality is what foreign economic policy is all about."[15] In regards to the specific means employed these can be grouped under four policy headings: commercial policy, foreign investment policy, foreign aid policy and balance of payments policy.

Some comments on the proposed framework. In the first place it is written by an economist for economists. Consequently many comparisons are made betwen economic and political behavior. As these comparisons are somewhat irritating for the political scientists, they have been omitted from the discussion.[16] By the same token, in view of its affinity with economic thought, the model of international political behavior has been lifted straight from the pages of political realism, whose deficiencies have already been examined. On one hand we find that the ascription of

---

[14]Ibid., p. 15.

[15]Ibid., p. 16.

[16]As an example, the comparison between oligopoly and state strategies (pp. 5-8) can only be made up to a point. While there are similarities the discussion misses the vital point that oligopolies follow their strategies within the limits imposed by a referee - the state - while the latter is constantly under the shadow of the *ultima ratio*.

foreign policy irrationality to internal politics, derived from institutionalized economic interests, begs the question of whose ratio one shall use as a standard when the actors disagree on ends. On the other hand we find the rather convenient definition of power under whose mantle economic relations can snugly fit.

But this need not detain us. As long as the influence of domestic politics and economic interests on foreign policy is taken into account it matters little whether this influence is deemed as rational or irrational. More important is Cohen's inadequate analysis of the interrelation between general foreign policy and foreign economic policy. Cohen's purpose is the development of a "systematic framework of analysis that would permit discussions of the allocation of scarce resources in support of power relations." Within the limitations outlined above, it fulfills the stated purpose. But the use of resources in support of power relations is only one facet of foreign policy. For unless one is to take cover under the all-embracing definition of power one will also have to consider the use of general foreign policy in support of foreign economic policy. In his subsequent discussion on United States foreign economic policy since 1945 Cohen states:

> In fact, this use of foreign economic policy represents a significant departure from the pattern that prevailed up to World War II, when it was more often diplomacy that was employed to promote America's commercial and financial interests abroad: today commerce and finance are manipulated in the interest of diplomacy.[17]

---

[17]Ibid., p. 29.

While this is an adequate description of the present relation between U.S. general foreign policy and foreign economic policy this framework is inadequate for the analysis of many countries, especially developing ones, and quite false in individual circumstances. One does not have to take the obvious case of developing countries who gear their general foreign policies to the maximization of foreign aid in order to prove the inadequacy of this aspect of Cohen's analytical framework. Take the recent settlement of the long standing dispute between Italy and Austria on the treatment of the German speaking inhabitants of the Italian Province of Alto Adige, formerly South Tyrol. As a result of the agreement Austria renounced all official interest in its lost province in return for which Italy removed its objection to the eventual inclusion of Austria into the Common Market. Italy's action is a clear case of foreign economic policy support of general foreign policy. But how does Austria's action fit into Cohen's analytical framework? In view of this conceptual difficulty we will not restrict ourselves to the analysis of the use of foreign economic policy in support of general foreign policy but, where relevant, examine the interrelation between both policies.

And now some adaptation of Cohen's framework for our purposes. He proposes the analysis of four economic policies in support of general foreign policy. These are commercial policy, foreign investment policy, foreign aid policy, and balance of payments policy. We will omit the last one but consider the effect of the first three economic policies on it. We will further restrict ourselves to the common features of the foreign economic policies of four groups of states: industrial and developing oil importers, oil exporters and states with Communist regimes. This classification roughly coincides with the membership of OECD, the "77" bloc of UNCTAD (minus oil exporters), OPEC and COMECON. But for reasons of space and organization, the commercial policy of OECD countries includes their energy policies, all three foreign economic policies of COMECON countries are lumped under commercial policy, and the foreign investment policy of OPEC countries includes their oil concession policies. Throughout the analysis emphasis is stressed on diversification in order to reduce dependence on others. But first an idea on what dependence means.

II. THE OIL BOYCOTT

The term boycott is as difficult to define as the term intervention. While most can agree on the clear-cut manifestations, these terms become more imprecise as one approaches the grey areas, between military occupation and high-pressure negotiations for the latter, and between direct stoppage and the unwillingness of resumption of commercial relations for the former. One can start describing the oil boycott by excluding acts between actively belligerent nations, as such acts are part of active war.[18] There are basically two types of oil boycotts;

---

[18]The terms *active* belligerents and *active* war are used advisedly, in order to exclude prolonged cease fires, a la Korea, various truce periods in the Middle East, etc.

the boycott of oil exporters' sales and the boycott of oil importers. We will consider both types.

The boycott of oil exporters' sales has so far been accomplished only by oil companies in direct retaliation for the nationalization of their assets. As the focus of this chapter is on the actions of governments we will describe the actions of the oil companies only as a background for analysis. History records three boycotted oil export nations: the Soviet Union, Mexico and Iran. In addition, at the time of writing, another boycott is taking place in Bolivia.

The boycott of Soviet oil exports started soon after the nationalization of the foreign oil companies by the Soviet Government. While most of the injured foreign oil companies agreed in principles to the boycott, its prime engine was Shell with Deterding at the helm. It will be remembered from Chapter II that the sale of Soviet oil in India by Standard of New York was the trigger that started the 1926 international price war among oil companies which, in turn, led to the formation of the inter-war cartel. The boycott of Soviet oil failed due to political and commercial reasons. After the resumption of trade relations between the Soviet Union and the rest of Europe, the problem was posed of what goods to trade. While the Soviets had a long shopping list their means of repayment was restricted to the proverbial furs and caviar and to primary commodities, notably oil. As the major European countries were eager to supply the large-scale Soviet demand for capital goods they agreed to the import of Soviet oil under the terms of bilateral treaties.

The political hurdles having been cleared Soviet oil began to capture a substantial share of the "western" countries.[19] The commercial success of Soviet oil was based on the presumed world shortage of oil at the beginning of the 1920's and, following that, on the normal Soviet commercial behavior which included forward integration into marketing facilities, limited price cutting, and market-sharing agreements with the majors in various countries.

The second oil exporters' sales boycott occurred in Mexico in 1938 after the nationalization of the oil companies which was the outcome of a ". . . revolution which the companies were, at that time, quite unwilling to accept."

> Their willingness to re-negotiate their concession arrangements and their largely extra-territorial and statutory positions - so extreme that some Companies had what amounted to private armies for protecting their lands - probably made nationlization inevitable.[20]

---

[19]"Mr E.P. Gurov, Chairman of Sojuzneftexport, the Russian oil export organization, claimed at the Beirut oil congress in 1960 that during the decade 1925-1935, Russia had supplied 14.3 per cent of all the oil imported by Western countries, and that during the peak years of 1930-1933 its share reached 19 per cent. About 14 per cent of Soviet production, he added, was exported during the thirties; its peak exports reached 30 per cent of its annual output." Hartshorne, op. cit., pp. 234-235.

[20]Peter R. Odell, "The Oil Industry in Latin America," in Penrose, op. cit., pp. 288-289.

In any event, the ensuing boycott by the major companies on Mexican oil, which also included ". . . her imports of goods and services to run the oil industry"[21] was all but complete, and " . . . eliminated [Mexican] oil exports - then running at a rate of over 3 million tons per annum - almost overnight."[22]

The Mexican nationalization of the oil companies and the ensuing oil boycott took place within a relatively inter-state political vacuum. Since the Mexican government agreed to pay adequate compensation and since the nationalization had little foreign policy implication it failed to affect U.S.-Mexican relations to any significant degree. The net effect of the boycott was to show the efficiency of that particular weapon and to force Mexico to concentrate on its growing internal demand.

The most significant boycott of an oil exporter's sales occurred after the nationalization of the Anglo-Iranian Company in 1951. This episode, which is commonly referred to as the Abadan dispute " . . . dragged on for two or more years, involving the International Court, the United Nations, and the intervention of the U.S., with the issue of compensation becoming central."[23] Both the Mexican and the Iranian nationalizations of their respective oil industries were caused by a combination of commercial and domestic political considerations, but

---

[21]Tanzer, op. cit., p. 32.
[22]Odell, op. cit., p. 291.
[23]Penrose, op. cit., p. 66.

while the former ran its course in a foreign policy vacuum the latter became from the outset enmeshed in international complications.

The commercial dispute evolved within the framework of the 50-50% round of negotiations which took place in the Middle East during the 1946-1950 period. The details are unimportant: the Iranians wanted more and the British were willing to bow to the inevitable. But the crucial factor was the large scale resentment of the preponderance of Anglo-Persian with its 51% British Government ownership in Iranian economic life, and of British influence in general. "To Iranians Anglo-Iranian and the British Government were virtually indistinguishable, and economic and political disputes inevitably interacted on each other."[24] Thus, the issue of higher returns from oil sales via nationalization became from the outset a confrontation between Iran and Great Britain.

With nationalization came a full-scale boycott which was adhered to by all the major oil companies and enforced on all tanker operators, and as a cause of the boycott "Iran was literally subjected to slow strangulation."[25] The reason why Iran was able to withstand this boycott for as long as it did " . . . was that the oil sector was a relatively autonomous island in the Iranian economy; . . . moreover, . . . oil revenues accounted for only about 12 percent of total government revenues."[26]

---

[24] Tugendhat, op. cit., p. 136.
[25] Ibid., p. 141.
[26] Tanzer, op. cit., p. 324.

The negotiation for the resumption of Iranian oil operations became a three-cornered affair between Great Britain, Iran and the United States. "The British Government counted heavily on the support of the United States, because of the latter's oil interest in Saudi Arabia, Iraq, and other areas of the Middle East. . ."[27]

> The Iranians calculated that . . . the United States would exercise pressure on the British to accept the inevitable. Their major assets, they felt, were American fear of Russian penetration into the Middle East in case of serious political difficulties with Iran, and the importance of Iranian oil to NATO and for the reconstruction and development of Western Europe.[28]

"The United States thus found itself playing the unenviable role of honest broker where each side expected the broker to take his part."[29] For reasons that need not detain us, the United States finally sided with Great Britain, going to the extent of making its foreign aid to Iran contingent on the latter's acceptance of a compromise with the former, and if one were to heed to substantive rumors, of staging a coup to oust the Mossadegh regime.

After the downfall of Mossadegh the new government lost no time in entering into substantive negotiations which resulted in the formation of the Iranian Consortium. The legacy of the Abadan dispute was a lesson *urbi et orbi* on the risks of dependence and confrontation. For the major

---

[27] Schwadran, *op. cit.*, p. 147.
[28] *Ibid.*, p. 150.
[29] *Ibid.*, p. 151.

companies it meant the beginning of diversification of oil production regions, a process that has been intensified as a result of subsequent Middle East crises.

> Another important result of the Iranian crisis was the warning it gave to the governments of other oil-producing countries of the danger of heeding those of its people demanding the nationalization of oil companies. . . . Thus, the dependence of the crude-oil producing countries upon the international majors was painfully brought home to their governments, and even in the tense aftermath of the war with Israel in 1967, none of the chief exporting countries attempted outright nationalization of the major crude-oil producing companies.[30]

The common feature of these three boycotts of oil exporters' sales is the fact that these were performed by oil companies as a result of the nationalization of their assets. The corrollary of this common feature is the purely commercial aspect of the boycott. With the increasing importance of inter-state oil trading one may well visualize a situation of government boycott of oil exporters' sales for foreign policy purposes. Thus, should the Algerians decide to depart considerably from their present foreign policies they would have to consider the effects of either a French or Soviet Government boycott of their oil sales. And just for argument's sake, let us assume that geology had been kinder to Rhodesia and transform it into an oil exporter. It takes little imagination to visualize a politically motivated international boycott of its oil sales instead of the present blockade of its oil import access route. With this speculation we can now turn to the boycott of oil importers.

---

[30] Penrose, op. cit., p. 67.

The boycott of oil importers can be classified according to the agent performing the boycott; these are either oil companies or governments. The actions of the former are of recent vintage and have occurred as a result of the post-World War II competition from Soviet oil in undeveloped countries. The tension arising from the pricing policies of the major international oil companies in oil importing undeveloped countries have already been discussed in the refinery section of Chapter I. It has been seen that the combination of oil company owned refineries, with crude supplied at posted prices without discounts, and of more advantageous Soviet crude offers, has produced some explosive situations. In Guinea the oil companies were forced to handle Soviet oil in their refineries, to which they reluctantly acquiesced. In India the Soviet offer of cheaper crude was used as a whipsaw to reduce the crude oil prices charged by the oil companies. But the same attempt in Ceylon was resisted by the oil companies. The result was the nationalization of their assets and the automatic application of the Hickenlooper amendment to U.S. foreign aid for Ceylon. But the most explosive situation, which resulted in a boycott took place in Cuba, to which we now turn our attention.

The story of the gradual transformation of Cuba under Castro from radical nationalistic socialism to radical left communism hardly needs any retelling. We will therefore concentrate on the early period of transition, where the situation among the three main actors -- the U.S., the U.S.S.R., and Cuba -- was still in a state of flux, and where Cuba was seeking the limits of permissability under the prevalent conditions.

"In 1960 Cuba was importing oil at a cost of about $80 million per year, which represented 10 percent of her total imports."[31] In February of that year, Cuba signed a trade agreement with the Soviet Union " . . . in which the Russians agreed to buy 5 million tons of sugar from Cuba in five years at approximately the world market price,"[32] in return for which the Soviets would supply, among other, what amounted to half of Cuba's oil needs. The Cubans asked the Jersey, Shell, and Texaco refineries to process this quantity of Soviet oil. The companies refused. Cuba nationalized ("intervened") the refineries. The companies boycotted Cuba.

A boycott of an oil importing country by oil companies is not a simple task. Barring a concurrence of all oil exporting countries - obviously not the case of the Soviet Union - it requires the acquiescence of all other oil companies who in turn must be able to control all means of transport. "The international majors were almost totally successful in getting other Western oil companies to agree to the boycott. In those few cases where they were not, the tanker boycott apparently did the trick. . . . Blocking the Soviet Union from obtaining tankers for shipping its oil to Cuba was a horse of a different color, however."[33]

The international majors control the bulk of the world tanker fleet. In order to implement a boycott they must be able to control the

---

[31] Tanzer, op. cit., p. 328.
[32] Ibid.
[33] Ibid., p. 337.

tankers on the spot market as well.

> In 1960 potentially there was available to the Russians more than twenty times the amount of tanker tonnage needed for the Cuban trade. . . . Put another way, for the boycott to be successful the oil companies would have to induce the owners of more than 95 percent of the theoretically available tanker tonnage to refrain from dealing with the Russians.[34]

Th main "inducement" of the oil companies is the black list for future charters, but "Working against the companies, however, is the highly competitive structure of the tanker market, in which a large number of operators each own a relatively small amount of total capacity."[35]

As borne out by subsequent events even the threat of the black list was insufficient to prevent the Soviet Union from chartering the required tanker tonnage for their trade with Havana. "Only the threat of armed force could stop the tankers, and Moscow had already warned that it was willing to reply with force."[36] And on this note we now turn to the oil boycott by governments.

Government boycotts of oil importing countries can be classified according to exporting, transit and other types of governments. Before turning to the first category, let us examine the others. Government oil boycotts by oil transit governments have been examined in some detail in the discussion on oil transportation in Chapter I. In the discussion it

---

[34]Ibid., pp. 337-338.

[35]Ibid., p. 338.

[36]Ibid., p. 343, quoting Harvey O'Connor, World Crisis in Oil (New York: Monthly Review Press, 1962), p. 262.

was pointed out that oil-pipeline-transit countries enjoy economic and political benefits from their geographical location, and that their bargaining power is equal to the cost of rerouting the pipelines through alternative countries plus the cost of lost production during the construction period of new pipelines. Discounting the two disruptions of the Suez Canal because of their blanket effect on all types of shipments, thus far the only transit government oil boycott occured in Syria during the 1956 crisis.[37] The pipeline sabotage was effective in the sense that it deprived Iraqi oil to Britain and France, "until the evacuation of Egypt by all the invading forces."[38] To what degree the Syrian action had effect on British and French post-Suez policies is a matter of conjecture.

Boycotts of oil importing governments by other than oil exporting or oil transit countries belong to the shadow area between war and peace, to the realm of the classical _casus belli_, the blockade. History records two such attempts, one of them still in progress: Italy and Rhodesia. Both attempts were failures. Both involved Great Britain and international organizations.

The background of the oil boycott of Italy is well known. For reasons best known to himself, Mussolini decided to acquire Abyssinia. Taken on its own merits, this would not have caused too great a stir, but

---

[37] For a rejection of the Syrian government claim that the sabotage of the IPC pipeline was performed by Syrian pipeline workers, see: Lenczowski, _op. cit._, p. 325.

[38] _Ibid._

"There was one awkward point: Abyssinia was a member of the League of Nations. . ."[39] and this involved the concept of collective security, ". . . a system where the British shouldered the burdens and others did the talking."[40] In any case, as the Italian invasion got under way the League opted for economic sanctions.

> The experts at Geneva reached the conclusion that the one economic sanction most certain to check Italy would be to refuse to supply it with petroleum and petroleum products . . .
>
> They estimated that a total embargo by all countries would result in an exhaustion of Italian stocks in something like three to three and one-half months.[41]

An effective oil embargo required first of all the concurrence of all the League members. This accomplished, it further required the compliance of powers outside of the League, notably of Germany and of the United States.

> This, too, was not serious. Hitler was playing for British friendship after the Anglo-German naval treaty; he was also delighted to see a dispute springing up between Italy and France. It was therefore worth his while to appear to be cooperating, unofficially, with the League of Nations . . . The United States, in the heyday of neutrality, could not take sides; but she forbade American trade with both combatants, and, as there was no American trade with Abyssinia, this was in fact a sanction against Italy.[42]

---

[39]A.J.P. Taylor, The Origins of the Second World War (New York: Faucet Publications, Inc., 1966), p. 89.

[40]Ibid., p. 91.

[41]Tanzer, op. cit., p. 326, citing Herbert Feis, "Oil for Italy: A Study" in his book, Seen from E.A.: Three International Episodes (New York: Alfred A. Knopf, 1947), pp. 305-306, emphases in the original.

[42]Taylor, op. cit., p. 91.

As a result, all Powers, excluding Italy's client states, agreed to the embargo and, consequently, the major oil companies ceased to import oil to Italy. But this was not sufficient as " . . . the temptations for a smaller company or even a group of individuals to provide oil for a boycotted country are great, owing to the potential enormous profitability for a small operator."[43] Temptations being created to be yielded to, Italy continued to import oil. The only effective way to stop this was by instituting a blockade, with its corresponding risks of war. A possible compromise was floated -- the Hoare-Laval plan -- but its premature publication left it on the vine. "The British government was still resolved not to risk war. They enquired of Mussolini whether he would object to his oil being cut off; when told that he would, they successfully resisted oil sanctions at Geneva."[44] All in all the League's actions were hardly effective. "Fifty-two nations had combined to resist aggression; all they accomplished was that Haile Selassie lost all his country instead of half."[45]

From one unsuccessful boycott we turn to another. In comparison with the straightforward case of aggression by Mussolini which represented a direct challenge to the moral authority of the League, the reasons for the United Nation's displeasure with Rhodesia are somewhat more obscure. Stripped of legal arguments the situation can be reduced to a few

---

[43]Tanzer, op. cit., p. 325.
[44]Taylor, op. cit., p. 95.
[45]Ibid., p. 95.

sentences. Rhodesia's caucasian population wishes to maintain the internal status quo. As it comprises a minority of the population it insists on a constitution that will perpetuate it in power, a prescription distasteful to the British Government. With the British Labour Party for once on the side of the angels, with one eye on the cohesion of the Commonwealth and the other on the Rhodesian order of battle, Britain brought its case against its rebellious colony to the United Nations.

On the assumption that Rhodesia's domestic policies are a provocation to its neighbors, who in turn will be forced to attack it, Rhodesia was declared to be a threat to the peace. Finding that neither Britain nor any significant Power was willing to live up to Sub-Sahara African principles by going to war with Rhodesia, these nations reluctantly agreed to lesser measures. Resolutions were passed, and a total trade embargo including oil, was placed on Rhodesia. In view of the alluring gains some oil runners tried to reach Rhodesia's pipeline terminal in Beira. In response to such an affront, and with nothing more serious to contend with than unarmed tankers, Britain instituted a full-scale blockade of the Mozambique Channel. With harbors blocked and honor safe, Britain and the international community rested its case. And if South Africa and Portugal persist in trading with Rhodesia, including with oil, let such deeds rest on their conscience.

Finally, we turn to the most sensitive of oil boycott types, to the one that gives to oil its unique role in international relations, to the boycott of oil importing countries by oil exporting governments.

For the past two decades this has meant in practice the actual or potential oil boycott by Middle East countries of the United States and of several countries of Western Europe. Let us briefly review the main factors. Geology has endowed many Middle East countries with abundant low cost oil and geography has blessed them with proximity to the main oil importers, whose industries and transport could grind to a halt for lack of oil. The bulk of the Middle East oil is located in Arab countries and produced by the international majors whose home countries are the United States and Great Britain. These two, together with Germany and for a time France, are de facto supporters of Israel, which most Arab states consider as their sworn enemy. Shake these ingredients with domestic instability, strain them through religious extremism, serve in a brittle glass of international tension, add a peel of fiery oratory and a dash of economic exploitation. Label the concoction political dynamite.

Looking back at the past two decades it would at first glance seem strange that in spite of all the potential friction sources there has been only one full-fledged oil boycott of Israel-supporting nations, and a short one at that. With the exception of the oil boycott of Israel, which is only part of its total boycott under the aegis of the Arab League, which has been countered by oil imports from the Soviet Union and later from Iran, and the previously described sabotage of the IPC pipeline in Syria, the only coordinated boycott occured during the aftermath of the 1967 Middle East war.

The 1967 oil boycott was actually a rather mild affair. It took the form of a notification to the various oil companies by Arab oil exporting countries to cease exporting oil to the countries on their black list: U.S.A., U.K. and Germany -- the latter for its sale of gas masks to Israel. This boycott was countered rather painlessly by U.S. oil exports and by a manipulation of offtake arrangements, to such an extent that Britain was in a position to refuse the generous offer to buy Soviet oil.

More serious was the blocking of the Suez Canal as this required (and still does) the rerouting the Persian Gulf oil via the Cape of Good Hope. But the situation was not as complex as the one encountered during the previous Suez crisis. "In 1956 more than 80 per cent of Western Europe's oil came from the Persian Gulf area, and had to pass through the Suez Canal. In 1967 the proportion was less than 60 per cent."[46]

One of the interesting side effects of the closing of the Suez Canal has been the cooperation between the Soviet and the British and French oil companies: Soviet oil exports to Japan originating from the Persian Gulf, and the same quantity of oil to Western Europe from the Black Sea. Another side effect has been the increased valorization, due to geographical advantage, of other than Persian Gulf oil, a fact which has not escaped Iraq's attention and for which it has demanded and received additional per barrel income. In other words, Iraq and to some

---

[46]Tugendhat, op. cit., p. 284.

extent Lybia have a direct interest in the closure of the Suez Canal, a fact which might partly explain the former's intransigence in regards to the present Middle East stalemate. The closing of the Canal has further benefitted South Africa by transforming some of its harbors into major bunkering stations, a fact not appreciated in some quarters. But perhaps the most important long-range factor has been the acceleration of the construction of supertankers which, for technical reasons, will deprive the Suez Canal of future tanker business.

Reverting to the 1967 oil boycott one should keep two factors in mind. On one hand the boycott took place under circumstances which the various countries, rightly or not, considered as extreme provocation. In any event it was the only weapon that they had and, in view of the previously discussed futility of nationalization, the only alternative to inaction. On the other hand, the boycott was of short duration. But this was not due to a belated return to a laissez-faire separation between commerce and foreign policy but rather to the realization that because of various factors the boycott was ineffective. Nevertheless, the boycott was to my knowledge without precedent in history insofar as it was an attempt by weak nations to impose their will on the foreign policy of Great Powers. That the attempt was made at all is a proof that it must be taken seriously. And the fact that it is, is presently examined. Let us begin with commercial policy.

III. COMMERCIAL POLICY

> Commercial policy represents the sum total of
> actions by the state intended to affect the extent,
> composition, and direction of its imports and exports
> of goods and services. These actions include not only
> the familiar direct interventions in international
> trade, such as tariffs, subsidies, quotas, exchange
> controls, official procurement policies, stade trading
> and the like, but also the many indirect interventions,
> ranging from domestic revenue taxes and pricing policies
> to sanitary regulations, advertising restrictions, and
> packaging requirements . . .[47]

Rather than to follow Cohen's generalizations on commercial policy according to his somewhat vague model, let us apply his findings with the proper modifications to our subject of interest. We begin with the industrial oil importers.

Industrial Oil Importers

Most industrial Western oil importing nations have the following common traits: a pluralistic society, a large private sector, a coal industry in need of protection, and a strong dislike of their dependence on oil exporters. Their commercial policies in regards to oil are part of their respective energy policies, which in turn are " . . . a balance between different and sometimes conflicting objectives."[48] Classifications being a matter of convenience, we will consider the main energy policy

---

[47]Cohen, op. cit., p. 20.

[48]Organization for Economic Cooperation and Development, Energy Policy, Problems and Objectives (Paris: OECD, 1966), p. 103.

objectives according to their effects on the state's oil import postures. One group of policies has direct or indirect, intended or tangential, short-term or long-term effects on the amount of oil imports. The other group of policies is specifically designed to cope with the vulnerability arising from the dependence on oil exporters. The first group of policies can be broadly classified in three categories: exploration and development of indigenous energy sources, all measures resulting in their protection, and all measures resulting in research and development of energy production and consumption technology.

A. Exploration and Development

"The mineral wealth of the three regions of the OECD and its various countries is far from uniform and policies differ accordingly. . . . Those countries with the resource potential have encouraged its development in the following principle ways."[49]

Tax Privileges. Fiscal measures are widely used to encourage indigenous exploration and production. In the United States the most important fiscal measures are the oil and gas depletion allowance,[50] the expensing of 'intangible expenses' including drilling and development costs against current income, and the limitation of income taxes to no more than 25 percent of capital gains realized from the sale of oil. One should keep in mind, however, that " . . . the benefits of the

---

[49]Ibid., p. 107.

[50]"Similar but lower allowances apply to coal and lignite, oil shale and fossile materials." Ibid., p. 108.

special tax treatment allowed by the Federal government to encourage
the discovery of oil and gas by the petroleum industry have in large
measure been captured by State and local governments."[51] Similar benefits are extended in Canada and in other OECD countries, but " . . .
because European oil and gas resources are small compared with North
America's the total impact of these incentives is correspondingly less."[52]

Direct Support to Exploration

> Some Member governments, Germany and France for
> example, have given direct financial support to
> exploration companies or given special loans or
> other financial aids for exploration. In the United
> States such financial aids, plus a guaranteed market,
> were used to stimulate exploration for uranium; and
> in Portugal, France and the United States, substantive
> amounts of the same commodity have been discovered as
> a result of exploration by government bodies.[53]

Other Measures. There are various government policies aimed
primarily at other objectives which nonetheless have the effect of promoting the exploration and/or development of indigenous resources. Thus, a
byproduct of oil and gas conservation policies is the diminishing of
waste. By the same token, "The terms of the leases of government land,
and of licenses to explore and exploit oil resources on land and sea,
are often designed to encourage exploration and development."[54] Other
such measures include financial incentives for the construction of hydroelectric plants for the purpose of advancing economic growth, regional
development, congressmen's political fortunes, etc.

---

[51]United States Department of the Interior, United States Petroleum Through 1980 (Washington, D.C.: 1968), p. 85.

[52]OECD, op. cit., p. 108.

[53]Ibid.

B. Protection

In addition to encouraging the development of competitive indigenous energy supplies, most of the energy producing countries of the OECD area have found it desirable in recent years to protect indigenous industries which have found it difficult to compete with imports.[55]

In practice this means the protection of European and Japanese coal from U.S. coal and Middle East oil, and the protection of U.S. oil from most OPEC area oil. The main protective measures against low cost oil are the following.

Import Restrictions and Duties. There are two types of **direct** trade barriers to oil: quantitative controls and tariffs. "Imports of petroleum are generally free of quantitative restraints in the major industrial countries, except the United States, France and Japan, which apply official controls."[56] The United States mandatory quota system is too well known to bear repetition. Besides, at the time of writing, there is a strong possibility that the system will be changed in favor of tariffs. Japan has a similar global quota system for the import of all petroleum products.

Imports of crude oil and finished products into France have been government-controlled since 1928. The marketing of petroleum products, especially gasoline, is subject to a quota system. Furthermore, refiners operating in France are obliged to take a certain percentage (at present 55 per cent) of their crude oil requirements from the franc zone, mainly from Algeria, at negotiated prices above the market level.[57]

---

[54]Ibid. [55]Ibid., p. 109.

[56]United Nations Conference on Trade and Development, Second Session, New Delhi. TD/97, Vol. II, Commodity Problems and Policies (New York: United Nations, 1968), p. 131.

We can disregard the modified quota system in Germany and turn to tariffs on crude and products. "Tariff restraints on imports of crude petroleum in the main developed countries are few, and where they exist, low."[58] The United States applies a tariff of 3-5 percent, Japan of 13 percent, while the common external tariff of the EEC has been fixed to zero. The purpose of these tariffs is mainly to bolster government revenue. On the other hand, "Import duties on refined products are more common and, in many cases, are relatively high."[59] Tariffs on oil products have the double purpose of encouraging domestic refineries and of protecting coal from the competition of fuel oil. But while tariffs are sufficient for the former special discriminatory taxation is additionally needed for the latter.

> Selective taxes on fuel oil are now levied in all major European coal producing countries, and even in countries without large indigenous coal industries there is a noticeably higher overall charge on fuel oil than on the other forms of energy competing in the fuel market. These taxes may be intended to produce revenue, to protect indigenous industry or to do both.[60]

"Taxation of non-substitutable products (e.g. oil for transport) usually has fiscal revenue as its sole objective and does little to protect

---

[57] Ibid.

[58] Ibid., p. 130.

[59] Ibid.

[60] OECD, op. cit., p. 110.

indigenous industries."[61] But while this is usually true in many countries, internal taxation can present an _indirect_ trade barrier to oil in others. "In 1967, taxes on gasoline represented as much as 77 per cent of the retail price in France and Italy, 72 per cent in the United Kingdom, 71 per cent in . . . Germany, and 70 per cent in Belgium."[62] Gasoline has been aptly described as "the tax collector's dream" insofar as there seems to be no short-term limit past which total gasoline tax receipts would decrease. But on the long run high taxation can reduce demand, " . . . as the price of gasoline influences the design and number of transport vehicles (e.g., small-cylindered versus big-cylindered automobiles)."[63]

Direct and Indirect Subsidies

Direct and indirect subsidies or other aids to indigenous industries are given by a number of countries. In Europe, the form in which subsidies and other aids are given to the different coal industries varies; examples are: wage subsidies and subsidies to miners' pension schemes . . . Sometimes subsidies are given to coal transport or to secondary fuel industries using coal; important aids have also been granted through recuperable advances, low-interest loans, state guaranteed loans and other financing facilities, either directly for coal production, or for its transport and use.[64]

---

[61]Ibid.

[62]UNCTAD, _op. cit._, p. 131.

[63]Ibid., p. 133.

[64]OECD, _op. cit._, p. 109.

The list of such subsidies is long. As usual, they are granted for a combination of reasons: social, regional, political -- coal miners are numerous and well organized -- and as a spur to rationalization, " . . . mostly in order to concentrate production in the most profitable mines or coal faces (negative rationalization) and to modernize production facilities (positive rationalization)."[65]

> Outside the coal sector, direct subsidies to the energy industries are rare. Germany, however, has recently replaced its system of customs duties to protect indigenous crude oil production by tapering subsidies, scheduled to end by 1970, on indigenous crude oil.[66]

Indirect Protection

> Sometimes protection is indirect. Examples are preferential transport rates or obligatory or tacitly observed restrictions of freedom of choice, such as the pressure brought to bear on electricity undertakings to purchase large quantities of indigenous coal under long-term contracts. Agreements of the latter type, for example, have played a part in protecting European and Japanese coal production.[67]

C. Research and Development

From the long-range perspective of this thesis perhaps the most significant energy policy of OECD countries, and especially of the United States, is the financing of research and development. "In most

---

[65] Ibid., p. 119.

[66] Ibid., p. 110.

[67] Ibid.

OECD countries, the greater part of research and development is undertaken by the private sector, but much of this is financed from public funds and governments also carry out research themselves."68

> The most impressive assistance towards research and development is that of the United States Government. The Federal authorities contribute to research bearing on all aspects of the discovery, production, conversion and use of solid and liquid fuels, to the study of certain aspects of energy transportation and to all sides of research and development of nuclear power, including construction of prototype nuclear reactors.69

While research and development is usually aimed primarily at other objectives, such as cost reduction via increased productivity and -- increasingly -- at the abatement of pollution, it has clear commercial policy implications insofar as it increases the availability and/or competitiveness of indigenous energy resources. "Research policy . . . is developing as a major responsibility of government, and its astute formulation may provide a key to the attainment of the long-term objectives in the energy field."70 One of these long-term objectives is security of supplies. "The emergence of nuclear energy as a serious competitor with conventional fuels, with the implications it has for the security of supplies, introduces a new dimension into energy policy . . ."71 But,

---

68Ibid., p. 118.

69Ibid., p. 119.

70Ibid., p. 136.

71Ibid., p. 142.

as concluded in Chapter III, the full effects of nuclear energy are still a long way off, while the oil boycott is a constant possibility. This then requires immediate preventive measures, to which we now turn our attention.

D. Preventive Measures

> Government policies have had an effect on the amount of energy entering into international trade. All measures of protection for indigenous industries - whether import quotas, duties, subsidies or other assistance - increase the demand for indigenous fuel at the expense of imports. . . . In spite of protection, international trade in energy has continued to increase both in countries with indigenous resources and those without. Oil is the main commodity involved."[72]

There are two complementary preventive measures against oil boycotts used by most of Europe's OECD member countries: diversification of sources of supply and stockpiling with its related measures. The latter have already been discussed in Chapter I under the heading of storage. It will be remembered that this involves all the physical measures that a nation undertakes on its territory to insure that it has sufficient oil and/or convertible energy sources to enable it to ". . . maintain its internal supply at a constant level, while oil supply sources are being switched."[73] Such measures include the stockpiling of oil and other energy sources in either conventional or unconventional

---

[72]Ibid., p. 128.

[73]OECD, Oil Today (1964), op. cit., p. 34.

storage, adequate domestic refining capacity, and the maintenance of a tanker surplus in order to compensate for longer distances, as in the case of the transport of Persian Gulf oil via the Cape of Good Hope instead of the Suez Canal.

But all these measures, including the ones designed to protect domestic sources of energy, are of no avail in the case of large dependence on one oil supplier. With this in mind we turn to the classical maneuver of commercial policy -- diversification of sources of supply. There are basically three policies or a combination of the three that are used by most of Europe's OECD members and Japan.

The first policy consists in the total reliance on the major international oil companies. As witnessed by the short-lived boycott of 1967, these companies have achieved a sufficient degree of diversification of sources of supply to counter all but a united front from all exporters, and a reliance on the major internationals would therefore be adequate for that purpose. Needless to say this does not mean exclusive reliance on one company and most of these countries have legislations with an anti-trust effect. But even so, for reasons that include national pride and balance of payments considerations, most governments prefer to take additional measures.

In the case of the second policy " . . . governments have taken steps to add to the number of suppliers by helping to create publicly owned or mixed national companies to find and produce oil abroad."[74]

---

[74]OECD, Energy Policy, op. cit., p. 111.

While this policy may result in additional security of supplies its main impact is the area of foreign investment policy, and is, therefore, discussed in that context.

The third and final policy alternative is government to government oil import agreements. Of these the most important are with the Soviet bloc countries, specially with the Soviet Union. "Although these agreements are usually concluded for other reasons, a certain diversification has been an incidental result."[75] The impact of these agreements is analyzed further in the text. But now let us turn to the second category of countries, to the oil importing undeveloped countries.

Developing Oil Importers

The commercial policies of developing oil importing nations are less elaborate than those of their industrial counterparts. Having little if any domestic energy to protect, no investments in oil exporting countries to speak of, and certainly no funds for energy research and development, their commercial policies can be reduced to a few alternatives.

In regards to the exploration and development of domestic oil and gas resources, and in view of the lack of private national capital for that purpose, the basic choice is between foreign private and domestic public funds. With the exception of some Latin American countries, notably of Mexico and Brazil and -- on-and-off -- Argentina and Bolivia, most countries have opted for foreign private capital -- in practice, for the

---

[75]Ibid.

international majors. Barring adequate domestic supplies the same choice is offered for the necessary down-stream investments. The alternatives between public domestic and private foreign refineries and, by extension storage and distribution systems, have been discussed at some length in the <u>refinery</u> section of Chapter I. As in the case of industrial nations, oil products are subject to heavy internal taxation for revenue purposes.

The problem of diversification of sources of supply is closely linked to the ownership of downstream facilities. Reliance on the international oil companies yields no discounts on posted prices but guarantees adequate supplies. Provided, of course, that the countries do not attempt to force the companies to lower their import prices or to refine Soviet oil, since this tends to lead to unpleasant consequences with the companies involved and, even worse, with their home countries. Within the present situation of oil and tanker surplus, state refinery ownership can bring net gains to the respective governments, as witnessed by Uruguay which has managed to secure its oil supplies since 1961 at discounts of some 20 to 25 per cent of the posted prices and at reduced freight rates.[76] Whatever Uruguay's present internal difficulties it is hardly likely to export them by attacking its neighbors. Its <u>raison d'être</u> being the one of the buffer state, neither of its giant neighbors will allow the other to annex it. This should then obviate any actual or threatened boycott by the home countries of the international majors with their control over

---

[76]Odell, "The Oil Industry in Latin America," in Penrose, <u>op. cit.</u>, p. 280.

tankers, as was the case during the 1965 India-Pakistan war.[77] This seems an appropriate transition for the the discussion of the commercial policies of the oil exporting nations.

Developing Oil Exporters

The commercial policies of most developing oil exporters share one trait with the export policies of the developing oil importers: problems arising from the overdependence on the export of one commodity.[78] This can be seen from the following statistics:

> Libya, Saudi Arabia and Kuwait derive virtually all their export earnings from petroleum. In the case of both Venezuela and Iraq petroleum represents more than 90 percent of total exports. The respective percentages for Iran, Algeria and Indonesia are 88, 61 and 38.[79]

In view of the fact that oil is the main source of foreign exchange of the developing oil exporters and that most of their oil is produced by foreign companies, oil exporters internal commercial oil policies are part of their concession policies, and discussed in conjunction with foreign investment policies. This then narrows the field to the oil exported by the national oil companies and to the control that the exporting governments can exercise over the destination of the oil exported by the foreign companies.

---

[77] See Tugendhat, op. cit., p. 287.

[78] But with one important difference. Oil is the single largest item on the import bill of many developing countries while tropical products are hardly a drain on the treasury of the oil exporters!

[79] UNCTAD, op. cit., p. 130.

Considering the fact that nearly all OPEC oil is produced by the international companies and by a few independents, and in view of the present and likely future glut of oil on the world market, the limited sales by the national oil companies can hardly be considered as a commercial policy maneuver for the purpose of diversification of purchasers. "These companies are the children, so to speak, of public opinion and are viewed as symbols of the national aspiration to gain a place in the sun on the international scene."[80] "The overriding aim, therefore, is to get into the market regardless of the consequences."[81] But this presents a problem " . . . since they have no desire to spend money abroad when there is so much to be done at home."[82]

> Without actual physical marketing outlets on the spot a national oil company would have to make a breakthrough so to speak, into an already existing pattern of supply and demand. In such circumstances, the only incentive that can prove attractive to the independent owner of the outlets or the final consumer is a further substantial reduction in the price . . . It is here in the final analysis that one finds a real conflict of interest; OPEC's desire for equitable (higher) *prices* and the eagerness of its national oil companies for *profits* whatever the price.[83]

In view of all the above and with few exceptions, notably Kuwait, the national oil companies try to export their oil without downstream investments and only in such areas where they do not compete with their

---

[80] Lufti, *op. cit.*, p. 37.

[81] *Ibid.*, p. 2.

[82] Tugendhat, *op. cit.*, p. 280.

[83] Lufti, *op. cit.*, p. 34. Italics in the original.

own oil sold by the international majors. This does not leave many alternatives. Kuwait is selling to Spain and, to its chagrin, has invested in a Rhodesian refinery. Iran is selling to Israel and South Africa, thereby losing popularity in some quarters. But the main market that seems to be opening up for large volumes of sales is Eastern Europe. This is examined later in the text.

While the OPEC member countries share the same concern in regard to their respective oil industries, they part ways when it comes to the use of the oil boycott in support of their foreign policies. In fact this aspect of commercial policy has so far remained the exclusive preserve of Arab oil exporters and transit countries. Insofar as Venezuela, Iran, and other non-Arab exporters are concerned they have had no quarrels with other countries to warrant an attempt to force the international majors to cut the oil supplies of their adversaries. In fact they have all profited in some degree from the various Arab boycotts. This also includes the Soviet Union, to which we now turn our attention.

The Soviet Bloc

The commercial oil policies of the Soviet bloc countries resemble OECD country policies insofar as they are tailored to the needs of industrial nations with by and large ailing coal industries, and are part of their respective energy policies. But due to differences in political regimes Soviet bloc countries use somewhat different administrative techniques. The energy policies of the Soviet bloc being by and large determined by

Soviet Union we will concentrate on the latter, with occasional reference to Eastern Europe.

The administration of Soviet oil exploration and development can dispense with the various measures used by the OECD countries. Simply stated, Gosplan -- the central planning board -- allocates funds and targets to the various ministries involved, and then awaits for the results. "Quite how these bureaucracies settle their demarcation disputes and apportion their overlapping responsibilities is a mystery to all except the Russian civil service."[84] In any case, somebody makes decisions and work gets done.

> Some of the key decisions of these unidentifiable managers, however, are made public from time to time. Over the past ten years Russia appears to have been expanding its crude oil output at about 12 per cent a year; there has been a considerable growth in refinery capacity and in pipeline transport. . . . By 1970, the Russians plan on an output of about 7 million barrels a day of oil and 240,000 million cubic meters of natural gas.[85]

In view of the fact that the Soviet Union has ample sources of energy, that it was tardy in changing its economy to an oil basis and that it has yet to enter the age of the automobile, it has had adequate oil supplies to cover its _total_ needs. "Like the United States, the Soviet Union prefers to use its own oil rather than anyone else's. Although imports from the nearby Middle East would be cheaper than domestic

---

[84]Tugendhat, _op. cit._, p. 247.
[85]Hartshorn, _op. cit._, p. 232.

supplies in many parts of the country, it, too, takes the view that it is unwise to rely on foreign oil."[86] Therefore, with the recent exception of supplies from Algeria, it does not import oil. And in view of the fact that its foreign trade is based on bilateral treaties it hardly needs the elaborate protectionist measures employed by OECD countries.

Before turning to Soviet commercial oil policies, a few words on their research and development policies. As opposed to most OECD countries, including the United States, the main reason for the spotty Soviet performance in that area is not lack of funds. "The Soviet leaders have never stinted in the supply of money to attain scientific and technological preeminence, even during the most trying of times."[87] In fact, the Soviet investments in science are quite impressive.

> For many years, the annual growth has been as high as 14 per cent and seldom under eight percent. . . Allotments in 1968 totaled some nine billion rubles as against some three billion ten years earlier. Several times as much is spent on science each year as on higher education. As for the future, the Soviet leaders . . . profess to remain convinced that investments in science yield greater dividends than other types of investment, and evidence an intent to double the present rate of support.[88]

The problem, as the Soviets are the first to recognize, is that ". . . the productivity of our [i.e., Soviet] scientists is approximately

---

[86] Tugendhat, op. cit., p. 244.

[87] Foy D. Kohler and Mose L. Harvey, "On Appraising Soviet Science and Technology," in Interplay, The Magazine of International Affairs, Nov. 1969, Vol. 3, No. 4, p. 21.

[88] Ibid.

two times lower than the productivity of the scientists of the USA."[89]
This is in large degree the result of " . . . one of the most basic features of the Soviet system: compartmentalization."[90]

Nevertheless, the allocation of large sums of money, coupled with "the great advantages that the socialist system provides in organizing science and industry," does produce significant results in high-priority areas. This is true not only for the space program and military hardware but is also for such energy equipment as hydro-electrical equipment, notably turbines, and extra high voltage transmission. At the time of writing the Russians seem to have a considerable lead in the area of sustained fusion reaction, with all its implications for the future.

While the poor performance of Soviet technology in many areas, notably consumer good industries, can partially be explained by reasons of their low priority, the stumbling block to increased efficiency is the application of scientific research to industry, a result of the cumbersome central planning system. A good example of the above is provided by the Soviet oil industry. While the Soviet drilling industry has proved its marks it has not produced enough refineries, tankers or pipelines to satisfy its post-World War II needs. Some of this was due to poor planning and some to bad timing, but for some strange reason the Soviets have up to now been unable to produce enough large-diameter steel pipes.

---

[89] Peter I. Kapitsa, Director of the Institute of Physical Problems of the Soviet Academy of Sciences, in a report to the U.S.S.R. Academy of Sciences on December 13, 1965, in Ibid., p. 20.

[90] Ibid., p. 19.

> In the 1950s it was their shortage of industrial raw materials, including steel pipe, without which their new oilfields could not be developed, that drove the Soviet Union into the oil export market. That is why their main sales campaign was in the developed countries outside of Western Europe and Japan.[91]

In contrast with the Inter-War period, the Soviets have not invested in downstream facilities in their export markets and have relied entirely on bilateral trade agreements. "A good deal of this oil, products and crude, has been sold at prices significantly lower than the generally scheduled prices in the markets outside. It would hardly have found purchasers if it had not been available at some discount."[92] Let us briefly examine these argreements according to the type of trading country. Soviet oil exports to Western Europe and Japan are guided mostly by commercial considerations. This sentiment is reciprocated. Needless to say, the Soviets realize the potential political implications of these oil sales but they must consider this a bonus. The reactions of European countries to this potential threat have been moderate. "Member nations of the European Economic Community have accepted an informal limit on their imports of Soviet oil, at 10 per cent of each country's imports."[93] In fact, as previously stated, the import of a moderate amount of Soviet oil -- say up to 20% -- provides Western Europe with a diversification of oil suppliers, and, therefore, with less dependence on Middle East oil. It is also good business.

---

[91]Tugendhat, op. cit., p. 249.

[92]Hartshorn, op. cit., p. 236.

[93]Ibid., p. 239.

In January 1967 The Economist estimated that over the previous few years Japan, Italy, and West Germany bought their Russian oil at an average price of between 7 and 8 rubles a ton. By contrast, Hungary had to pay 20 rubles a ton, Czechoslovakia 18, Poland 17, Bulgaria 16, and East Germany 15.94

At first glance this disparity in prices would seem to indicate a ruthless overcharging of a captive market based on the convincing argument of at least 20 Soviet divisions on East European soil. But appearances are deceptive, even to the parties involved. The Soviets insist that they can import better quality equipment from Western Europe for the same quantity of raw material, including oil, that they export to Eastern Europe, a charge that the latter deny. The question of who overcharges whom is quite moot in view of the administrative basis of the Comecon pricing system. A story has it that a Polish trade official once said at a Comecon meeting that after the inevitable world revolution the Socialist world would have to preserve at least one Capitalist state in order to maintain a system of world prices. Si non e vero é ben trovato!

"In the underdeveloped countries the political motivation behind Russia's willingness to enter into barter deals involving oil is much greater."95 While the political composition of Soviet oil depends on the individual circumstances, "Many Russian offers in the late fifties seemed neatly selected to offer the major companies the maximum of embarassment . . ."96 The temptations of the developing countries to purchase oil in

---

94Tugendhat, op. cit., pp. 251-252.

95Ibid., p. 250.

96Hartshorn, op. cit., p. 236.

return for whatever commodity they happen to produce have already been examined. So has the predicament of the major companies. An agreement to match the Soviet prices or to refine Soviet oil does not only reduce profits but can also produce a domino effect leading to a _de facto_ "most favored _importing_ country" situation, while a refusal may lead to nationalization. In the first case the Soviets get the credit for reducing the country's import bill; in the alternative case they may fish in troubled waters, specially if the nationalization results in a worsening of relations with the home countries of the major companies. In any case, a classical example of economic warfare.

But as of late, the Soviet "oil offensive" into developing countries has subsided. This may have been caused by many reasons: the Soviets may have improved their accounting methods and realized the true cost of their oil, or they may have grown tired of getting saddled with more cocoa than they can use;[97] their reverses in Ghana and Indonesia have certainly brought home the notion that political influence in unstable countries is a sometime thing. Probably a combination of all the above. But most important has been the change in their oil reserves, to which we now turn our attention.

---

[97] Because of the widespread resentment that this practice has caused, the Soviets have ceased to use the classical expedient of re-export. An example would be the barter of Soviet arms and equipment for Burmese rice or Egyptian cotton, which would be promptly resold on the world market for hard currency. This achieved two effects: dependence on further long-term Soviet import agreements due to the loss of previous export markets, and what amounted to the export of otherwise unsellable Soviet equipment for hard currency!

Just when things were getting to be interesting, geology reared its ugly head. Not that the Soviet Union is likely to incur oil shortages but its old production fields near Baku and from the Volga-Ural area from which the bulk of its production comes are getting exhausted while the new ones are located in Western Siberia and Central Asia.-- expensive sites from which to pump oil. On the other hand, if the Russians are going to have their automobile age and unless they go into a crash research program on electrical battery cars, the Soviet Union will have to consume most of its oil. "It its output achieves the target of 630m. tons laid down in the provisional plan for 1980 its own needs are likely to account for 560 m. tons."[98] The combination of less yet -- because of transport -- more expensive oil for export is going to require a realignment of priorities. Sales to Western Europe and Japan will continue to be of importance to the Soviet economy; this will then leave less mischief oil for the underdeveloped countries and very little for Eastern Europe, let alone for China.

> By 1980 it is estimated that Eastern Europe will probably require about 170 m. tons of oil a year, and its local production from the small fields in Rumania, Hungary, and Poland will amount to a mere 30 m. tons. Russia will be quite unable to fill the gap.[99]

The reorientation of trade patterns takes time, especially when it is of the magnitude of about 75% of oil imports, and as a result the Soviet Union has recommended to its East European allies "to sound out

---

[98]Tungendhat, op. cit., p. 253.

[99]Ibid.

the possibility of purchasing this raw material from other, especially developing states."[100] Having fulfilled its proletarian obligation of purchasing Soviet oil at a time of surplus, Eastern Europe will be allowed to purchase Middle East and North African oil at a time of relative Soviet oil scarcity, thereby opening up the first major market for the national oil companies of the exporting countries.

This reorientation of foreign trade is well in keeping with the new Soviet policy of natural gas import. "Two import contracts have so far been signed, with Iran and Afghanistan respectively. The one with Iran is much larger, and provides a first rate example of how successfully the Russians can integrate their commercial and foreign policies."[101] The problem with associated natural gas, it will be remembered, is that it can be only partially recirculated into the reservoir and that is liquification is rather expensive. Accordingly, without a pipeline, the production of oil results in the waste of gas. This being the case in Iran, the Shah could hardly resist the Soviet offer of a long-term purchase of a yearly maximum of 10 billion cubic meters of gas. For the Soviets this means reduced transportation costs; for the Shah, profits from previous waste and a lessening of dependence on the international oil companies. "But his relations with the Soviet Union are now too close to ever risk a really serious rupture in anything short of a major crisis."[102]

---

[100] New York Times, Nov. 24, 1969, quoting Vlastimil Plechac, deputy chairman of the Czechoslovak Committee for the Chemical Industry in an interview with the press agency CTK.

[101] Tugendhat, op. cit., p. 254.

[102] Ibid., p. 255.

The Soviet purchase of Middle East natural gas may well reduce its transportation costs in some areas, but there still remains the problem of what to do with its own natural gas from the new Siberian oil fields. As usual, Eastern Europe may be relied upon to accept a reasonable amount but this is a far cry from an adequate solution. Accordingly, after years of negotiations and having found the precise political timing, the Soviet Union has achieved a major breakthrough. In December 1969 it signed an agreement with West Germany whereby it will supply the latter with Siberian gas for 20 years at a maximum rate of 3 billion cubic meters per year. The pipeline which, needless to say, will be manufactured by West Germany, will be laid through Czechoslovakia, "depriving the East Germans of any revenue or political concessions that they might have obtained from the transaction."[103]

During the same month the Soviet Union signed a similar agreement with Italy for a 20 year delivery of natural gas at a maximum rate of 6 billion cubic meters per year. The pipeline will be made from Italian pipe, will be laid from the Austrian Czech border to Tarvisio in Northern Italy and will be built by an Italian-Austrian-Soviet consortium.[104] Further negotiations are in progress to extend the pipeline to France, "thereby inextricably linking the Soviet gas fields into the Western European distribution systems. . . If the scheme comes off it will be the biggest step forward in the integration of the Russian and Western economies since the Revolution of 1917."[105] This should give food for thought.

---

[103]New York Times, December 1, 1969.

IV. FOREIGN INVESTMENT POLICY

Foreign investment policy represents the sum total of actions by the state intended to affect the extent, composition, and direction of private direct and portfolio investments both by residents abroad and by foreigners domestically. These actions include principally taxes and administrative regulations of various kinds, but also monetary policy to a certain extent. Their purpose is to manipulate the stock as well as the flow of private foreign investments in such a way as to jointly maximize the state's national power and current income.[106]

On a somewhat abstract level and in regards to the maximization of current income there is a certain convergence of interests between capital exporting and capital importing countries. "Otherwise, we would hardly expect to find such large flows of private capital as we do observe in the world today."[107] For the former, foreign investments may provide higher returns on capital than domestically and long-term support for the balance of payments. For the latter, it represents a supplement to domestic savings and short-term support for the balance of payments.

As would be expected, while there is a certain convergence of national interests in regards to the maximization of current income there tends to be conflicting interests in the maximization of national power.

---

[104]*New York Times*, Dec. 11, 1969.

[105]Tugendhal, op. cit., p. 255.

[106]Cohen, op. cit., pp. 23-24.

[107]Ibid., p. 24.

"A basic motive of foreign investment is to minimize the danger of stoppages in the capital-exporting country's foreign trade by increasing the reliability of markets for exports and sources of supply for imports."[108] This basic motive of foreign investment policy usually conflicts with the security of capital-importing countries unless they are willing to be dependent on the capital-exporting countries. "Thus a capital-importing country will often, on security grounds, seek to constrain the flow of private investments from abroad below the level that might otherwise be dictated by purely commercial considerations."[109] As in the case of commercial policy, the basic strategy for the lessening of dependence on others is diversification. Having sketched the essentials of this model we can now adapt it to our requirements. In so doing we will consider the foreign investment policies concerning oil of OECD and OPEC member countries.

Industrial Oil Importers

In the discussion on the commercial policies of OECD member countries it was stated that most of these countries use a combination of three policies in order to diversify their sources of oil supply. These policies are reliance on the international majors, state trading with oil exporting countries, including with the Soviet Union, and the creation of publicly-owned or mixed national companies to find and produce oil abroad, the latter belonging more clearly to the realm of foreign investment policy. This is presently examined.

---

[108]Ibid.

[109]Ibid., p. 25.

302.

Let us recapitulate the comparative advantages of reliance on the international majors by most OECD countries. From an economic standpoint, and except for the U.S., the U.K., and to some extent the Netherlands (the home countries of the international majors), most of these countries benefit from the heavy downstream investments by these companies, for insofar as these investments derive from foreign sources they represent an addition to their respective domestic savings. But the price for this benefit is a larger foreign exchange outlay than warranted by present market conditions and a loss of fiscal revenues on downstream operations because of accounting manipulations. From a political standpoint the far-flung operations of these companies insure an adequate diversification of sources of supply, and thus enchance the security of the various importing countries. But the price to be paid for the security from potential stoppages by the exporting countries is a dependence on the international majors and on their respective home countries.

How do the publicly-owned or mixed national companies of OECD member countries producing oil in exporting countries fit in these calculations? For the sake of simplicity let us assume that their oil production costs, including taxes to the exporting countries, are equal to the costs of the international majors, and examine the comparative economic and political advantages for the respective OECD member countries.

From an economic standpoint, under present market conditions, the major advantage for the respective countries is a reduction in their foreign exchange outlay, approximately equivalent to the prevalent discount rates on crude -- say 20% -- that the international majors offer to independent refiners. The balance of payments advantage, however, must be balanced against the investment, usually from domestic sources, in refinery, storage and distribution functions. The main exception to this oversimplified scheme is provided by France where "The international companies are obliged to buy 55 per cent of their crude oil requirements from franc-area producers at a specially high price, and as their own reserves in the area are inadequate this in effect means giving a subsidy to the French producers."[110]

National companies may provide further economic benefits to the various OECD countries by increasing the competition among various companies in their respective domestic markets, but not to the point of depriving coal of further markets. National companies can also serve the purpose of the entreprise témoin, of the test-case of the operational economics of the international majors.

The relative political advantages of the national companies depend on individual circumstances, but one may venture some generalizations. National companies may provide an additional measure of security of supplies by lessening the dependence on the flux of relations between exporting countries and the international majors and their home countries, and between the latter and the other OECD countries. In the first case, the history of the last twenty years in the Middle East has shown that occasional advantages of having a production source independent of the international majors. But even good relations with the exporting countries

---

[110]Tugendhat, op. cit., p. 288.

is no guarantee against stoppages incurred by domestic groups.

> In the aftermath of the Arab-Israel war the Petroleum Worker's Union in some of the Persian Gulf ports refused to allow tankers going to Japan to be loaded, although it is difficult to imagine a country farther removed from the scene of the trouble. The ban lasted only a few days, but it made a deep impression on the Japanese Government.[111]

Perhaps more important, national companies may be granted concessions in countries where the international majors are not represented, or only nominally so. Up to now the only example in this category is provided by French companies in Algeria. Turning to the second case, to the lessening of dependence on the international majors and on their home governments one must differentiate between actions aimed at third countries and those aimed at the OECD countries themselves. "For example, an American oil Company in Japan (Caltex) was reported to have refused to supply lubricating oil to a British ship at a Japanese port because the ship was on a time charter to the Russians and was carrying Cuban sugar to Siberia."[112]

While national oil companies of OECD countries may reduce the likelihood of such politically motivated harassment against third parties on their own territory they may *in extremis* blunt a boycott against themselves. One does not have to go to the inter-war period for examples, such as the boycott against Japan that precipitated the attack on Pearl Harbor. "In his admirable book about the Suez Affair, Hugh Thomas reports that a senior, though anonymous, British Cabinet Minister told him in the

---

[111] Tugendhat, op. cit., p. 287.

[112] Penrose, op. cit., p. 269.

aftermath of the invasion in 1956 and 1957 the United States threatened to curtail Britain's oil supplies in order to influence her policies."[113] It surely is a matter of speculation but one tends to wonder whether de Gaulle would have had such a free hand in shaping his foreign policies if France had been largely dependent on the international majors for oil supplies.

Having examined the relative economic and political advantages of national oil companies to their respective OECD home countries, we can now turn our attention to the relations between the major international oil companies and their home countries. The difference between national and international oil companies reflects itself in the different interests of their respective home countries. National OECD oil companies are financed primarily by public and/or private domestic capital. These companies explore and produce oil in one or more exporting countries and import the produced oil for home consumption. Therefore these companies conform to Cohen's model and analysis. But the major international oil companies do not follow this pattern for the following reasons.

The international majors are essentially middlemen deriving a large part of their profits outside of their respective home countries. As an example, and if one excludes its operations in the United States, Jersey Standard produces in fourteen countries, refines in thirty-seven and sells in more than a hundred under its brand name of Esso.[114] This

---

[113]Tugendhat, op. cit., pp. 287-288, quoting Hugh Thomas, The Suez Affair, Weidenfeld and Nicolson.

[114]Ibid., p. 181.

then raises the question of the nationality of the capital invested by the international majors. Like any corporation they have access to three main sources of capital: equity, debentures, and reinvested earnings.

In regards to equity, "The distribution of the nationality of the shareholders of a firm is not easy to discover . . ."[115] but even apart from known cases, such as the half ownership of British Petroleum by the British Government, it is reasonably safe to assume that a majority - say over 66% - of the share of the international majors are owned by nationals of their respective home countries. The known exception to this statement is provided by Shell which is owned: "39 percent by United Kingdom citizens, 19 percent by Americans, 18 percent by Dutch, 12 percent by French, 10 percent by Swiss, and 2 percent by others."[116]

The distribution of the nationality of debenture holders is even more difficult to determine than the nationality of shareholders, especially when the bonds are floated on the European capital market with its predilection for anonymity. But the role of debentures, while growing, is not significant because ". . . in the fifteen years to the beginning of 1966 the industry financed only 5 per cent of its capital expenditure via the capital markets in London, Zürich, New York, Franfurt, and other financial centers."[117] In fact, the oil companies are largely self-financed, or put

---

[115]Penrose, op. cit., p. 41.

[116]Ibid.

[117]Tugendhat, op. cit., p. 260.

in another way, financed by price levels that " . . . make the consumer pay for their capital expenditure whether he likes it or not."[118] And in view of their global sales network one can hardly ascribe their capital to the savings of the economies of their respective home countries.

> Thus the legal ownership, which governs the statistical classifications of national income accounts, obscures one important aspect of economic reality, and may lead the unwary to assume that the foreign investment of the firm is a contribution of goods and services to foreign economies which is made possible by savings generated within the U.S. geographical economy. In fact, the term 'United States foreign investment' refers only to ownership of the investment, and ownership implies that both income and capital may be at any time transferred to the U.S. (unless prohibited by the foreign government). The specifically American contribution to the foreign economies may be little more than the willingness of the American parent to refrain from transferring to the U.S. profits that its subsidiaries have made in foreign markets.[119]

Reverting to the comparison between OECD national and major international oil companies, one notes that the former import oil for domestic consumption and are financed by national public and/or private capital and by forced savings from national consumers, while the latter's global sales networks enable them to be financed predominantly by forced savings from international consumers. Accordingly, and with the exception of the oil imported into their respective countries, the economic interest of the home countries of the international majors in the foreign operation of their respective companies is based mainly on their impact on the balance of payments.

---

[118]Ibid.

[119]Penrose, op. cit., pp. 42-43.

Accurate up-to-date figures on the effect of the international majors on the balance of payments of their respective home countries are hard to come by. Professor Penrose has examined the annual reports of the international majors and other pertinent material. From her cited work (published in 1968) one gathers that while all the major U.S. companies reported that the operations brought a net credit to the U.S. balance of payments, only Standard Jersey was more specific. "In 1964 the Company's net contribution to the U.S. balance of payments was reported to exceed a third of a billion dollars (Annual Report, p. 5) with still greater 'improvement' reported in 1965."[120] Royal Dutch/Shell reported on a basis of 'net credit,' i.e., taking no account of oil imports to the UK and the Netherlands. "It reported that its 'net credit' to the UK balance of payments averaged over ₤90 million in the six years 1961-6, and to the Dutch balance of payments apparently some ₤36 million in the decade 1956-66."[121] In regards to British Petroleum, "In 1965 net earnings on overseas trade less capital investment were reported as ₤61 million . . ."[122]

Overall "oil balance of payment" statistics are even more difficult to come by than from individual companies. In his cited work (published in 1969) Tanzer states that "The most definitive published study for the United States is that of the Chase Manhattan Bank."[123]

---

[120] Ibid., p. 101.
[121] Ibid., p. 108.
[122] Ibid., p. 116.
[123] Tanzer, op. cit., p. 45.

This source notes that in 1964 the net positive balance of the U.S. oil industry as a whole was $0.5 billion. But according to the Chase Study "one must distinguish between affiliate and nonaffiliate transactions."[124]

> What emerges from this is a true picture of the even greater impact of the investment in foreign affiliates of the integrated U.S. international oil companies. Thus, while the total contribution of the oil industry to the balance of payments as a whole in 1964 was $467 million, this was made up of a positive contribution on the balance with overseas affiliates of $743 million, which more than offset a $277 million deficit for the petroleum balance with other nonaffiliate foreigners. That is to say, without the overseas affiliates of the international oil companies, the United States balance of payments deficit in 1964 would have been one-quarter greater than it was.[125]

The impact of the international majors on the balance of payments of their respective home countries is even more striking for the U.K. According to another source ". . . over the 1955-64 period oil imports of the two British internationals . . . amounted to $5.3 billion, which was more than offset by their earnings of $6.0 billion . . ."[126] But in view of the fact that the U.K. has no oil reserves and that it would have had to import the oil regardless, " . . . the real contribution of the two companies to Great Britain's balance of payments is the $6 billion which they earned for the country. . ."[127] Considering

---

[124]Ibid., p. 46, citing the Chase Manhattan Bank, Balance of Payments of the Petroleum Industry (New York: October 1966), p. 11.

[125]Ibid.

[126]Ibid., p. 48, citing The Economist (London), May 15, 1966.

[127]Ibid.

that over the period 1961-1965 Britain had an annual trade deficit of $2-$3 billion[128] one may well state that this "invisible" income from international oil profits has nearly made the difference "between a shaky solvency and bankruptcy."[129]

These statistics, however sketchy, give an indication of the importance of the foreign operations of the international majors to their respective home countries. These statistics also explain why the U.S. and the U.K. provide diplomatic support for the protection and advancement of these foreign operations. Accordingly, and to the extent that the support of these foreign operations does not clash with other foreign policy objectives, one may roughly equate the foreign interests of the international majors with the national interest of their home countries. This includes not only commercial interests such as foreign taxation and investment policies but also strategic considerations such as diversification of supply, pipeline and refinery locations, etc. The latter also coincide with their respective national interests as oil importers.

So far we have discussed the diplomatic support of commercial operations. We now must consider the reverse, the use of the foreign operations of the international majors to support the foreign policies of their home countries. With the exception of both world wars and of

---

[128] International Monetary Fund, International Bank for Reconstruction and Development, Direction of Trade. A Supplement to International Financial Statistics, Annual 1961-1965.

[129] Tanzer, op. cit., p. 48.

both boycotts inspired by international organizations, Britain has made scant use of its political control over two international majors for the purpose of denying oil to an importer. Britain maintains the sensible point of view that it does not have to like a nation or its policies in order to trade with it.

The United States has had fewer compunctions to use its control over foreign affiliates of U.S. international majors in support of its foreign policy objectives. Apart from the already mentioned boycotts against its foes during both world wars, and against Italy and Rhodesia, and apart from the alleged threatened boycotts against Britain during the Suez Crisis and against India and Pakistan during their 1965 war, the United States has maintained an almost total trade embargo against China, North Korea, North Vietnam and Cuba. It is doubtful whether the oil boycott against its present Asian opponents has more than a symbolic value because these countries are unlikely to wish to become dependent on the United States for this strategic commodity. On the other hand, the boycott of the Organization of American States against Cuba has been partially successful in its stated aims. True, Cuba has had to redirect its trade and has become totally dependent on the Soviet Union for its oil. But under the present circumstances, with the Soviet Union acting as a dampener on Cuba's export of revolution to Latin America, Cuba's dependence on the Soviet Union is not without some merits for the OAS member countries.

Developing Oil Exporters

The foreign investment policies of oil exporting countries operate within a different framework than the one encountered in industrial oil importing nations. The governments of industrial nations view oil within the context of commercial, foreign investment, domestic energy and industrial policies; accordingly, their total policies concerning oil are best examined within the context of these often overlapping and sometimes conflicting policies. But for oil exporting countries these policies are better subsumed under the heading of concession policy -- a euphemism for semi-diplomatic negotiations.

> It is appropriate to treat the economic relations between the oil Companies and the developing countries laregly in terms of 'bargaining' and of balance of power, rather than in terms of ordinary trade and commercial competition, because the conditions in which trade takes place, as well as the nature of the fiscal and other financial arrangements arrived at are frequently renegotiated when changes take place in external circumstances or in the effective bargaining position of the governments.[130]

The history of oil concessions has been interspaced in the narrative and analysis of Chapter II. We will therefore recapitulate only the essential aspects in order to determine how concession policies can be used to support foreign policy objectives. The pre-World War II concessions in the Middle East were granted for periods ranging between 60 and 75 years, covered very large areas -- occasionally entire countries -- and allowed little if any government control in the operations

---

[130] Penrose, op. cit., p. 260.

[131] Parra, op. cit., p. 40.

313.

of the oil companies. In return for these terms, the latter paid the governments a flat royalty and were exempt from further taxation. After long negotiations and/or unilateral actions by governments, acreage was rescinded and the government take increased. The post-1948 concessions include a greater variety of terms than in the old concessions, terms with varying degrees of advantages for the oil exporting countries. But the impact of these new concessions on the international oil industry structure has so far been negligible.

Within the present circumstances, how can oil exporting nations manipulate their oil concessions in order to maximize their current income and their national power? The optimal policy should be one that provides the maximum benefit to the economy, lessens the dependence on the international majors, and increases the security of outlets. Are these objectives compatible?

Let us first consider the economic objectives. There is not much that can be done about the established concessions, most of them granted to the international majors. Barring an OPEC-wide nationalization or strictly enforced proration program, only scant additional benefits can be extracted from these companies within the present buyer's market. Not for lack of will or imagination: schemes to "optimize" national revenues litter the desks of petroleum ministries. These schemes include higher income taxes and royalties, "the imposition of an additional tax over and above the normal income tax, which would limit a company's rate of return on investment to a certain maximum,"[130] limitations on profit repatriation, equity participation, infrastructure and social projects, higher wages and benefits for national workers and employees, etc., etc. But all the items on this shopping list have the same effect: reduce the profit margin of the companies and driving them to the irreducible minimum past which operations cease to be profitable.

Barring additional substantial benefits from measures that amount to an increased per barrel take, governments may take measures to the effect of decreasing their per barrel take in order to gain additional revenues from increased volume. But, as OPEC admits, this is not without its perils, as the increased volume would have to come at the expense of other oil exporting countries.

> In view of the overall inelasticity of demand, it would indeed be suicidal for any major producing area to encourage development of its own resources over an extended period of time by low taxes at the expense of other areas. This form of competition in taxes would in the long run only serve to drive down Government revenue for all.[132]

So far the governments of OPEC have resisted the temptation of reducing taxes in order to increase volume. At least openly. But this does not mean that they do not wish to increase their respective shares of the world market, no matter at whose expense. "Iran often openly expresses dissatisfaction with the limited growth of its output in comparison with such states as Kuwait, which has 15-20 per cent more output and no more than an eightieth of its population."[133] But Kuwait does not agree with this line of reasoning.

---

[132] *Taxation Economics in Crude Petroleum*, a paper presented at the Fifth Arab Petroleum Congress, p. 16, cited in Penrose, op. cit., p. 206.

[133] Hartshorn, op. cit., p. 327.

While the freedom of action of OPEC governments in respect to established concessions is limited because of world market considerations they enjoy a greater leeway in the bargaining for new concessions. "In some countries, where the monopoly enjoyed by the majors had for long been considered prima facie evidence that they were 'exploiting' the countries, independents seeking concessions were not only welcomed but sometimes favored."[134] But experience showed that these independents became active price cutters because of their lack of vertical integration. This gave many OPEC countries second thoughts about the relative merits of monopoly.

But the fact remains that for security reasons the international majors are looking for oil outside of the OPEC area, especially outside of Arab countries, while, in order to gain access to promising geological formations, the new concessionaires are willing to enter into new types of agreements, including equity sharing and service contracts. Accordingly, and with the exception of Venezuela, OPEC countries continue to grant concessions to these lesser companies, even at the expense of some acreage held by the international majors. This also coincides with the security objectives of OPEC countries, to which we now turn our attention.

The granting of concessions that result in oil production to companies other than the international majors provides the OPEC countries with a diversification that enhances their respective bargaining power and security of outlets. But while American independents only dilute the

---

[134]Penrose, op. cit., p. 209.

oligopsony of the international majors, companies of other nationality, whether public, private or mixed-capital, lessen the dependence on the United States and Great Britain. In addition, the threat of withholding new concessions may be used to influence the energy policies of OECD countries.[135]

Nevertheless, the oil produced from these concessions is negligible in comparison with the total produced by the international majors. And most OPEC countries are faced with a situation where " . . . a very small handful of foreign firms have it in their power by their own decision to damage the national economy . . .; the fact that they can use it only in extremis . . . alleviates only a little the prevailing resentment."[136] Within the present situation, what are the alternatives? Let us examine some of the possibilities.

One measure adopted in varying degrees by most oil exporting countries is the diversification of export commodities. Accordingly, Venezuela has increased its export of iron ore, Algeria of wine, and Indonesia of various tropical commodities. But the Middle East oil exporters do not have alternative commodities. Thus, like all other OPEC

---

[135] The Conference, having noted that the policies of certain industrialized countries may tend to artificially depress petroleum prices in international markets; . . . recommends that Member Countries should not grant any new oil rights of any nature whatsoever to companies whose home countries pursue such policies; . . .
Resolution XVII.94 at the Sixteenth Conference of the Organization of Petroleum Exporting Countries, Vienna, June 1968.

[136] Penrose, op. cit., p. 250.

countries, they either export natural gas or prepare to do so in the near future. Even more attractive is the development of a domestic petrochemical industry which can increase domestic consumption of oil and gas, save foreign exchange on previously imported petrochemical products, and provide the basis for an export of industrial products. Within a few years most oil exporting countries "should be producing sizeable quantities of fertilizer, chemicals, plastics, and other materials derived from oil."[137]

Another measure that can be used to reduce the dependence on the international majors is state trading. Some of the variations on this theme have already been discussed. These include concessions to public OECD companies and the trading by OPEC national oil companies. But state trading in significant amounts requires government-to-government agreements. One such type of agreement would leave the assets of the international majors in oil exporting countries intact, but curtail their freedom of operations by means of bilateral or multilateral treaties. The latter coincides with the desires of the E.E.C. Commission.[138]

A much more radical possibility would be to tax all foreign oil companies out of production profits, thus obviating the bother of nationalization. This would require prior assurance of guaranteed markets and prices. But unless the prohibitive taxation and the market guarantee

---

[137]Tugendhat, op. cit., p. 278.

[138]See: Commission des Communautés Européennes, Première Orientation pour une Politique Energétique Communautaire, (Communication de la Commission au Conseil), COM (68)1040, Brussels, 18 December 1968, proposal No. 21.

were simultaneous and multilateral the lone exporting country attempting such a step would only change one form of dependence for another. For in the face of the ensuing boycott by the international majors it would find itself at the commercial and political mercy of the importing country.

While the relative importance of state trading is bound to increase in the future, especially in view of the opening markets in Eastern Europe, OPEC countries have made little headway in their attempts to bypass the international majors in their respective markets. This is due partly to the strenuous resistance of the international majors (and of their home countries) and partly to the lack of enthusiasm for such arrangements by most OECD countries. As a result of this impasse OPEC countries are trying to reduce their dependence on the international majors by exercising a larger degree of control over their operations via equity participation.

There are various possible types of equity participation. The first that comes to mind is an oil exporting country purchasing stock in a major oil company operating on its soil. Such an action is within the means of Kuwait who, over the years, could well have acquired a controlling interest in Gulf. But such a massive capital export from a developing to an industrial country is even more expensive and less practical than forward integration by OPEC national oil companies outside of their home countries.

Another variant of equity participation is one which by various means results in a higher take for the government but which leaves the essential control over the operations in the hands of the foreign company. This is the general effect of "service contracts," where the company puts the money for the exploration and is repaid only if oil is found, but where the payment - in oil - is "for services rendered, not by right of ownership."[139] But apart from semantic gratification the government enjoys no greater control than before. Such is the experience of Iran with the Consortium.

> But if the effect of participation by the government is to put in it in a position to control the entire use of the profits arising in the local company or to influence the level of production, real difficulties arise . . . as . . . joint ownership of the major crude-oil producing affiliates of the international Companies could give rise to pressures from governments not only to extend the partnership to all their vertically integrated activities but also to relate downstream investment directly to particular sources of oil. This would severely hamper, if not destroy, the Companies' cherished 'flexibility' of operations if applied to their major sources of crude oil, while pressures from the governments for an increased share of profit would be in no way abated.[140]

With so much at stake the issue of participation leading to control over operations has become a highly political issue. But for the time being, and within the present market conditions, OPEC countries

---

[139]Tugendhat, op. cit., p. 279.

[140]Penrose, op. cit., p. 214.

are unable to bring about this redistribution of power. In view of the lack of cooperation from the major oil importing nations, OPEC can not parlay physical control over crude oil into financial and operational control over production levels in their own countries or over transportation, refining and distribution in third countries. Discounting drastic and concerted actions OPEC countries will have to wait for a change in their respective bargaining positions with the international majors. Such a change might occur if, for reasons discussed in Chapter II, the prices of crude oil and products " . . . reach a point at which the Companies feel compelled to ask for re-negotiation of the existing financial arrangements . . ."[141] It might well be ironic if OPEC's failure in maintaining the price of oil became instrumental in its achieving a greater degree of control over it. On the other hand, a continued drop in prices might prove to be the proverbial last straw and lead to a different situation altogether.

> If the trends described above -- increased taxes, government intervention, and falling market prices -- continue, the oil Companies may well find it desirable to explore means of selling their major producing affiliates to the governments of the oil-producing countries and substituting some form of contractual relationship.[142]

Such a 'voluntary nationalization' would prove to be more than unfavorable for the oil-producing countries. They would achieve the political goal of state trading but in view of the continued oligopsony

---

[141] Ibid., p. 210.
[142] Ibid., p. 255.

of the international majors based on their downstream investment, the OPEC countries would be as dependent on them as before. More, in fact, if one considers the loss of bargaining power caused by the departure of "hostages". The economic consequences would be even worse. With no more investments to protect in OPEC countries, the international majors would have little incentive to maintain the high price of crude oil. Like any other manufacturer in similar circumstances they would use their market power to reduce the price of their primary commodity while attempting to maintain the price of their manufactured goods. And part of these new profits could be spent on increased insurance premiums: stockpiling in Western Europe and intensified exploration outside of the Middle East. In any event, such a transformation of the international oil industry structure could not come about without the full unleashing of competitive forces, an unpleasant prospect for a seller in a buyer's market.

V. FOREIGN AID POLICY

Foreign aid policy represents the sum total of actions by the state intended to affect the extent, composition, and direction of foreign public assistance given or received. Like other foreign economic policies, its purpose is to maximize jointly the state's national power and current income. Its analysis is fairly straightforward since the main economic gains all accrue to the recipient while the main political benefits all accrue to the donor. Thus the recipient will seek to maximize the benefits of foreign aid obtained at least cost in terms of dependence on other states, while the donor will seek to maximize its influence abroad at least cost in terms of foregone alternatives on public capital.[143]

---

[143]Cohen, op. cit., p. 26.

Let us examine the applications of this novel tool of economic diplomacy to our subject. OECD member countries give foreign aid on both bilateral and multilateral bases, but for economic and political reasons the bulk is given in bilateral form. Bilateral foreign aid provides the donor countries with varying degrees of influence over the policies of recipient governments. This influence is used, inter allia, to support the respective oil interests of OECD countries. Up to now no OECD country has provided foreign aid funds for the development of state-owned oil industries in underdeveloped countries. This is not only in line with a development doctrine that emphasizes infrastructure and development projects but it also coincides with the interests of the home countries of oil companies with international operations.

The influence that OECD countries may exercise on the oil policies of developing countries is not restricted to this essentially negative aspect. Insofar as foreign aid is tied to efficient allocation of resources, donor countries may withhold foreign aid should they deem the economic policies of recipient governments unsound. Such unsound policies include the allocation of scarce public resources for which foreign private capital is available.

But perhaps the most direct use of foreign aid in support of foreign economic objectives is the protection of foreign investments. While many examples may be cited of implicit connections by OECD countries between foreign aid and foreign investment only the United States has made this connection statutory, in the form of the Hickenlooper Amendment. Since the formation of OPEC and with the recent exception of Indonesia, no major oil exporting country has received any significant U.S. bilateral aid, thus depriving the U.S. of a leverage to support its international oil companies. But this leverage can be effective in other instances. In Ceylon it accomplished the purpose of adequate compensation for the nationalization of oil company assets. Nevertheless, the cessation of foreign aid is a rather blunt instrument that should be used with caution as it inevitably brings a worsening of governmental relations. This is why the U.S. has refrained until now to invoke the Hickenlooper Amendment in Peru and Bolivia after the recent nationalizations without adequate compensation of IPC -- a subsidiary of Jersey Standard -- and of Gulf, respectively.

Bilateral foreign aid is not the only way that donor countries can influence the oil policies of recipient governments; while less direct, multilateral aid achieves the same purpose. The degree and manner of control varies with the type of international organizations. In the case of the World Bank and of its affiliated organizations, the control by OECD countries is straightforward insofar as the voting power of member governments in this institution is determined by their subscribed capital. "Thus, while the Bank has been a major lender to underdeveloped countries in the energy field, with more than one-third of its loans going to that sector, it has never lent any money for petroleum refining or exploration in underdeveloped countries."[144] It should be noted that

---

[144]Tanzer, op. cit., pp. 92-93.

the Bank's refusal of loans for petroleum activities is not based on the defense of U.S. and U.K. interests but rather on the economic assessment that such activities are better left in the hands of foreign private firms with the necessary capital and expertise, but in any event the result is the same.

But in international organizations based on the "one country one vote" system OECD countries, notably the United States, must rely on other measures to protect their interests. Among these measures the most effective is the withholding of funds for projects not of their liking or, failing that, the threat of withholding funds for other programs, the so-called financial veto. The best proof of the effectiveness of this veto is that in spite of numerous protests to the contrary, the United Nations has been virtually silent and inactive on the question of petroleum economics. Such an intricate interplay between national and international economics and politics provides a commentary on our times. But in order to comment on it properly we will change our variables and focus of analysis. This is the task of the next chapter.

CHAPTER V

OIL AND INTERNATIONAL RELATIONS

I. INTRODUCTION

In the introduction to Part II we examined two types of interpretations of the relations between economics and politics in international relations. Marxism and political realism were set as parameters, one emphasizing economics and society and the other politics and state. Both were found inadequate for the analysis of the role of oil in international relations because of their epistemology based on tautology leading to a perception of constant ends. Having noted the difficulties arising from the fact that the subject of our inquiry is one of many variables with multiple causation whose relations and causation differ according to levels of abstraction and conceptualization, and in time, we opted for a two-pronged approach. The first approach was on the basis of focus of interest, the oil boycott, a phenomenon that lies at the juncture of general foreign policy and of foreign economic policy. This led to the adaptation for our purpose of Cohen's conceptual framework. We now turn to the other approach.

Let us first consider general foreign policy and foreign economic policy in a static model encompassing the relations between man, state and the international system according to an ideal-type of their respective operational milieus. Thus we can start with an Aristotelian position which considers political man as the "integrator" of political man in the narrow sense, economic man, religious, moral and ideological man, etc., who pursues his often conflicting interests according to his available means within the limits imposed by the interests and operational milieus of the various collectivities to which he belongs.

By shifting the focus to the nation-state we observe groups of political men representing various constellations of interest and power, acting within the operational milieu of the unit's "objective factors" (geography, technological level, economic resources, populations, military potential), pattern of power and political culture.[1] While the operational milieu of the nation-state varies according to the individual units, its essence is a sense of community with various degrees of cohesion, and the formation of the volonté générale with the political unit as the referee and as the holder of the domestic order based on the preponderance of organized violence against individuals and groups. Part of this volonté générale concerns the relations of the nation-state with others, and takes the form of foreign policy which in turn operates within the milieu of international anarchy with its systemic risk of war and corresponding behavior.

---

[1] For a discussion of the terms pattern of power and political culture, see: Hoffman, Contemporary Theory in International Relations, op. cit., pp. 180-181.

Seen from this perspective we can thus consider general foreign policy and foreign economic policy as the product of the interaction between constellations of domestic interests and power within the national operational milieu and of constellations of state interests and power within the international operational milieu. Such a perspective fits the flexibility requirements of the model by allowing the observer to analyze the behavior of actors which he deems as important for the subject of his inquiry according to a *Zweckrational* scheme which distinguishes both constellations of interest and power, and operational milieus.

This is the model used in the previous chapter. As long as the main variables comprising the international system are constant, this model allows concentration on foreign policy without taking systemic causes into account. But as the time span of analysis increases so do systemic variations. In view of the fact that this chapter will analyse the period from the beginning of the oil industry to its presumed depolitization we will have to use a dynamic model of the international system that includes its systemic variations. Hoffmann's model of the international system fulfills this requirement.

> An international system is a pattern of relations among the basic units of world politics, characterized by the scope of the objectives pursued by those units and of the tasks performed among them, as well as by the means used to achieve these goals. This pattern is largely determined by the structure of the world, the nature of the forces that operate across or within the major units, and the capabilities, patterns of power, and political cultures of those units.[2]

---

[2]Hoffmann, *The State of War*, op. cit., p. 90.

According to Hoffmann, a new international system emerges when, by using stakes of conflict as the main criteria, there is a new answer to the following questions: what *are* the units in potential conflict, what *can* the units do to one another, and what do the units *want* to do to one another?[3] From the beginning of the oil industry to the present time one can distinguish the following international systems: the pre-World War I period, the Inter-war period, and the post World War II period. These international systems will be presently examined in light of their influence on the various oil industry structures.

II. THE PRE-WORLD WAR I INTERNATIONAL SYSTEM

The oil industry began its existence during a peaceful decade of a relatively harmonious century. Let us examine how this political atmosphere influenced the industry's development. For this we must first turn to the world order established at the Vienna Conference after the defeat of Napoleon, whose " . . . ambitions produced the first modern instance of total power politics, based on an ideological inspiration and waged by total domestic and international means."[4] The victors at Vienna decided to return to the balance of power system that had prevailed since the treaty of Westphalia, a system that included France as one of

---

[3]Ibid., pp. 92-93.

[4]Ibid., p. 102.

the main actors. In this system Austria and Russia wanted to "include a formula for domestic order in its concept of legitimacy and dispose of means of enforcement against the rise of liberal and nationalist forces."[5]

> But this was not to happen, for it soon appeared that a voluntary system of cooperation was too weak to control developments within nations that a previous balancing system had already been powerless to prevent. In other words, so extensive a community could not be created by superstructural means alone. The failure of the Holy Alliance proved that an effective new balancing system could be obtained only through a return to moderation, not through an ambitious extension in scope and means."[6]

There were two main factors that contributed to moderation in international relations: a fluid balance of power and effective transnational ties. "The structure of the world was marked by a double hierarchy: the distinction between a civilized core and a frontier and, within the core, between the small and large states."[7] The political framework of this structure was defined by the Concert: "the instrument of the society of the major powers whereby they could supervise small states and control the individual ambitions of each member . . ."[8] Within this multipolar system the latter function required a fluidity of alignments which in turn "required neutrality towards regimes as well."[9] This

---

[5]Ibid., p. 103.
[6]Ibid.
[7]Ibid.
[8]Ibid., p. 104.
[9]Ibid.

neutrality was reinforced by transnational ties among national elites. "The _Internationale_ of diplomats allowed for a concensus on the rules of the game. Although regimes were far from identical, the limited state developed everywhere."10

The return of moderation in scope and means in the relations among states that was achieved at Versailles was a prerequisite for the development of the limited state. Within the limited state and with the masses barred from the political process by means of property qualifications on voting rights, the middle class was able to resume the primitive accumulation phase of the industrial revolution without considerations of social cost. With political power in the hands of the middle class the _laissez-faire_ state could well advocate the harmony of interests doctrine that underlined the separation between political and economic power, between state and society. The _laissez-faire_ state was indeed, as per the Communist Manifesto, "a committee for managing the common affairs of the whole bourgeoisie." In any event the _laissez-faire_ state transformed the international economic society.

> This new international economic society was built on the fact of progressive expansion and on the theory of _laissez-faire_. The expansion of Europe consisting both in a startling increase in the population and production of Europe itself and in an unprecedentedly rapid dissemination of the population, products and material civilization of Europe throughout other continents, created the fundamental change from the static order and outlook of the 18th Century to the dynamic order and outlook of the 19th.11

10_Ibid_.

Having shed the mercantilist policies of previous centuries, the 19th century instituted an unprecedented freedom of trade and of migration. The latter coincided with the interests of middle class governments who "concerned with the importance of cheap and abundant labor . . . had been under no political compulsion to give prior consideration to the wage-levels and standards of living of their own workers."[12] But freedom of trade clearly did not coincide with the short-term interests of some governments. Yet this period transformed a series of local economies into a world market. Why?

Some reasons have already been stated. These include the moderation in international relations leading to the limited state, and the increase in production and population leading to a dynamic outlook that " . . . opened a 'depoliticized' zone for free trade and for the free establishment of aliens. . ."[13] To this, one should add the improvements in rail and sea transportation that lowered the cost of delivering products to the final consumers. But the fact that economic nationalism did not appear during this period, or at least only in muted form can be attributed to gaps in theory of political economy.

---

[11]Edward Hallet Carr, Nationalism and After (London: MacMillan & Co., Lt., 1945), pp. 11-12.

[12]Ibid., p. 22.

[13]Hoffmann, op. cit., p. 105.

The success of this 19th-century compromise
between a closely-knit world economic system and
unqualified recognition of the political diversity
and independence of nations was rendered possible
by two subtle and valuable pieces of make-believe
which were largely unconscious and contained sufficient elements of reality to make them plausible.
These two salutory illusions were, first, that the
world economic system was truly international, and
second, that the economic and political systems were
entirely separate and operated independently of each
other.[14]

The first illusion was based on the assumption that the interna-

tional economic system "was not an artificial creation of man but part of

an order of nature."[15] The unparalleled expansion in production seemed

to prove that the 'invisible hand' hailed by Adam Smith harmonized all

interests, foreign as well as domestic. But seen in retrospect, "progres-

sive expansion was the product not of the principle of universal free

trade (which was never applied, and whose application would have been

intolerable) but of the open British market."[16] It was this "apparently

insatiable" import and entrepôt market for "all consumable commodities

... which made the so-called international system work."[17]

The international system, simple in its conception
but infinitely complex in its technique, called into
being a delicate and powerful financial machine whose
seat was in the city of London. The corollary of an
international commodity market was an international
discount market, and an international market for shipping
freights, an international insurance market and, finally

---

[14]Carr, op. cit., p. 13.

[15]Ibid.

[16]Ibid., p. 14. My emphasis.

[17]Ibid.

an international capital market. All this required and depended on the effective maintenance of a single international monetary standard into which national currencies were exchangeable at fixed rates; and this in turn presupposed a central control over the currency policies of the different national units, enforced by the potential sanction of a refusal to deal in "unsound" currencies.18

Thus the progressive expansion of production -- the pillar of the 19th century doctrine of harmony of interests -- was absorbed in large part by the British import and <u>entrepôt</u> market while the stability of the international economic system was maintained by the 'pure' gold standard that insured total convertibility of all currencies, and controlled by Lombard Streed in a fashion that was "autocratic, without appeal and completely effective."19

"The second illusion which secured acceptance of the 19th century world order sprang from the formal divorce between political and economic order."20 As long as the international economic order was thought to be a part of an order of nature there was no awareness of the submission to British economic authority. "This economic authority was a political fact of the first importance; and the British economic power of which it was a function was inseparably bound up with the political power conferred by the uncontested supremacy of the British navy."21 But while the

---

18<u>Ibid.</u>, p. 16.
19<u>Ibid.</u>
20<u>Ibid.</u>
21<u>Ibid.</u>

international economic order was based on British hegemony, it functioned impartially, in the sense that it did not serve British interests "in any narrow or exclusive sense . . . (as) . . . the commerce of the world was a British concern."[22] This may partially explain why most nations accepted "the discipline of a supreme external arbiter of their economic destinies in the disguise of a law of nature."[23]

Having examined the world order we can now turn to its influence on the development of the oil industry. But first a few words on the influence of the world order on the development of the United States. On the political side the Concert of the Great Powers gave the United States the isolation from world affairs required for its economic development. With Britain as the balancer of the international system and with the British navy as the enforcer of the Monroe Doctrine the United States was able to dispense with the economic and social costs required for the maintenance of a large military establishment. On the economic side the advantages were equally compelling. The U.S. was a net importer of capital. It imposed prohibitive tariffs yet benefitted from free access to foreign markets. The vast flow of immigrants provided the U.S. with abundant manpower and a growing internal market. And for reasons that need not detain us universal suffrage failed to hamper the laissez-faire development of industrialization.

---

[22] Ibid.

[23] Ibid., p. 17.

When Drake dug his first well in 1859 he was lucky to find oil. The industry that he pioneered was lucky to find an environment which boosted its development. The Anglo-Saxon legal framework produced the "law of capture" that encouraged maximum production and competition. This led on one hand to the displacement of coal and shale oil for the production of kerosene, and of vegetable and animal oils for the production of lubricants. On the other hand this competition led to a situation of a multitude of marginal producers vying for a share of a growing but fragmented market. But the political framework of laissez-faire which reflected the popular belief that 'only the weak are good because they are not strong enough to be bad' permitted Rockefeller to build his oil empire without serious government interference.

Turning to the international scene Standard Oil found no government impediments to its capture of foreign markets. Tariffs were applied but without discrimination of import source. As few countries had indigenous oil supply and as fuel oil was too expensive to compete with coal, tariffs had the main purpose of raising fiscal revenue and had little effect on total demand for oil. Government intervention in the oil industry was reserved to control over oil fields. "Until shortly before the outbreak of the First World War Dutch interests . . . and the Dutch Government, blocked Standard's efforts to obtain concessions to explore for oil in the Dutch East Indies. And in Burma and India,

British interests . . . successfully opposed similar efforts."[24] Needless to say this brought diplomatic representation from the U.S. government under the "open door" doctrine.

The period between 1880 and the First World War was the era of classical imperialism where "the basic problem of international relations was who should cut up the victims."[25] The dismemberment of Africa provided for some authors the proof of the relation between capitalism at the stage of monopoly and the inevitability of contradictions leading to war. But the explanation of the conquest of Africa is somewhat more pedestrian.

> Imperial conquest remained, in the minds of the European statesmen, the sign of greatness. Europe was at peace, the Western Hemisphere protected by the Monroe Doctrine. One took what remained to be taken, and the unwritten law of compensation, which the diplomacy of cabinets obeyed, obliged the states to claim in turn a share of a continent which all could have safely ignored.[26]

This did not mean that "once the conquest was made the colonists did not dominate and exploit the vanquished, as all conquerors have done throughout the ages."[27] Outside of colonies which governments tended to

---

[24]Penrose, op. cit., p. 55.

[25]W.L. Langer, The Diplomacy of Imperialism, ii. p. 797, quoted in E.H. Carr, The Twenty Year's Crisis, 1919-1939 (New York: Harper & Row, Harper Torchbooks, 1964), p. 49.

[26]Aron, op. cit., pp. 266-67.

[27]Ibid., p. 266.

maintain as a special reserve for their nationals there was comparatively little government direction in the areas of foreign trade and investment. Despite preferential trade between colonies and their respective mother countries the volume of trade among industrial nations was higher than between industrial and non-industrial nations.28 In other words, trade followed the flag only when it was found to be profitable. When required, private companies sought assistance from their respective home governments, but foreign trade and investment was usually conducted without regard to what today is considered as national interest. "During the pre-World War I period the actions of the oil monopolies constantly interfered with the prevalent political groupings and the going process of international relations."29

Laissez-faire allowed the formation of Standard Oil, whose commercial practices forced its remaining domestic and foreign competitors into vertical integration. The relatively free flow of goods and capital across national boundaries without concern to international politics allowed the development of the international oil industry. International vertical integration of oil companies was greeted by Lenin as "socialization of production."30 According to Foursenko, "whatever the arguments of the advocates of monopolies, the fact remains that the very nature of monopoly capital inevitably gives rise to such phenomena

---

28This seemingly obvious statement is overlooked by proponents of the theory of imperialism.

29Foursenko, op. cit., p. 457. My translation.

30. ... when these products are distributed according to a

as cosmopolitanism and antinational policies."[31] But limited government intervention in the economy and transnational ties are the symptoms of a homogeneous international society. And that society was shattered by World War I.

III. THE INTER-WAR INTERNATIONAL SYSTEM

Great wars are traditional catalysts of domestic and international change. World War I accentuated the trends that began as a result of the unification of Germany, trends that were to transform the international system. "Germany lay at the point where the lines of pressure of Continental high politics intersected . . . Integral German unification . . . was hardly attainable without revolutionary shifts of power . . ."[32] This soon became evident with the outcome of the Franco-German War and the annexation of Alsace and Lorraine; this outcome in turn created a lasting focus of tension among two Great Powers, thus destroying the fluidity of alliances required for the proper balancing of the system. Feeling constrained by its borders, Germany turned to the sea and began the construction of an impressive battle fleet, thus striking at the heart of

---

single plan among tens and hundreds of millions of consumers (the distribution of oil in America and Germany by the American "oil trust") -- then it becomes evident that we have socialization of production. . . V.I. Lenin, Imperialism, The Highest Stage of Capitalism (Peking: Foreign Language Press, 1965), p. 154.

[31]Foursenko, op. cit., p. 457.

[32]Ludwig Dehio, The Precarious Balance. Four Centuries of the European Power Struggle, Charles Fullman translator (New York: Random House, Vintage Books, 1962), p. 216.

British power. "In the end, Britain made up her mind to view Germany as Enemy Number One, an opponent of the order of sixteenth-century Spain or of France in the seventeenth, eighteenth, and nineteenth centuries -- in other words, as a power seeking supremacy."[33] This challenge required among others an increase and improvement of the British fleet, and led to the change-over from coal to oil as "The use of oil made it possible in every type of vessel to have more gun power and more speed for less size and cost."[34]

Germany's new strength and ambitions upset the fluidity of alliances and was instrumental in the formation of two rigid blocs of states. But other forces affecting the homogeneity of the international system were also at work. In 1879 Germany imposed "the first 'scientific' tariff - a piece of economic manipulation in the interests of national policy."[35] Economic affairs were again becoming important in international politics. Franchise was gradually extended to the masses, and as a result ". . . horizontal links between major powers were progressively weakened by the rise of mass nationalism . . ."[36] The combination of these factors would not necessarily have crystalized into a new international system had World War I taken the course envisaged in war games, such as the Schliefen Plan. A relatively short war would hardly have transformed

---

[33] Ibid., p. 235.

[34] Winston Churchill, The World Crisis,1911-14, cited in Tugendhat, op. cit., p. 67.

[35] Carr, Nationalism and After, p. 17.

[36] Hoffmann, op. cit., p. 108.

domestic political systems. General staffs tend to plan their battles on the experiences derived from previous wars. But instead of the "blitzkrieg" of the Franco-German war they should rather have built their expectations on the model of the U.S. civil war.

World War I was the first "total war" of modern times waged by industrial countries using mass conscription. No country was prepared for the carnage that it produced. Whatever the initial war aims the fury of battle transformed them into a desire for total victory which in turn required the harnessing of total means. War ceased to be a limited quarrel among governments and became a fight to the finish among nations. The means ate up the ends.

The use of mass conscription by countries with mass suffrage " . . . suggested to those in power than henceforth war must have a meaning for those who risked their lives in it."[37] For the first time in modern times hatred of the enemy became an officially encouraged policy. For the first time since the Napoleonic wars total blockade on enemy trade was instituted. "For four years, the Central Powers were compelled to depend exclusively on their own resources, and to realize in spite of themselves Fichte's ideal of the Closed Commercial State."[38]

> After 1914 both personal relations and commercial transactions, direct or indirect, with enemy citizens became a criminal offense; and for the first time in the history of modern war enemy private property was confiscated - a devastating blow at the foundations of laissez-faire society and bourgeois civilization.[39]

---

[37] Aron, op. cit., p. 102.

[38] Carr, The Twenty Year's Crisis, p. 123.

[39] Carr, Nationalism and After, p. 27.

"Semi-homogenous in 1914, the European system had become irremediably heterogenous by 1917 as a result of the fury of the struggle and the Wester powers' need to justify their determination to win decisively."[40] The first link to snap under the tension was the Tsarist regime; the ensuing dismemberment of the Russian Empire was an omen of things to come. The American entry into the war tipped the balance decisively in favor of the Allies, who -- according to Lord Curzon -- proceeded to float to victory on a wave of oil. And so came the armistice.

> The Germans duly acknowledged defeat; in return
> - and almost without realizing it - the Allies acknowledged the German government. . . . The armistice settled the question of Germany unity so far as the first World war was concerned. The Habsburg Monarchy and the Ottoman Empire vanished. The German Reich remained in existence.[41]

The treaty of Versailles altered the map of Europe. It dismembered Empires in the name of national self-determination, lumped nationalities together in the name of national viability and, in the case of Austria, prohibited self-determination in the interest of power politics. Emperors had given way to democracies who, per definition, would pursue peaceful foreign policies. But there was a problem: "however democratic and pacific Germany might become, she remained by far the greatest Power on the continent of Europe; with the disappearance of Russia, more so than before."[42] But this lay in the future; in the meantime Germany was prostrated and had to pay.

---

[40]Aron, op. cit., p. 103.

[41]Taylor, op. cit., p. 26.

[42]Ibid.

The treaty of Versailles was the first major peace treaty imposed by democracies responsible to an electorate. And the masses wanted tribute from the vanquished, a tribute to be paid in goods and money. Among others, the victors wanted coal, but "Germany could only execute the coal demands of the treaty by abandoning the bulk of her industries and returning to the status of an agricultural country."[43] As Keynes wrote in 1920: "Taken in their entirety, the economic clauses of the treaty are comprehensive, and little has been overlooked which might impoverish Germany now or obstruct her development in the future."[44]

Bismarck had set a modern precedent by extracting some reparation from the French after the Franco-German war, but British and French politicians had promised their constituents that Germany, in the famous phrase, would be squeezed till the pips squeaked. Such was the magnitude of reparations demanded that by signing the treaty Germany "in effect engaged herself to hand over to the Allies the whole of her surplus production in perpetuity."[45]

---

[43] John Maynard Keynes, "The Peace of Versailles," *Everybody's Magazine*, LXIII (September 1920), cited in Ivo J. Lederer, Ed., *The Versailes Settlement. Was it Foredoomed to Failure?* (Boston: C.D. Heath and Company, 1965), p. 45.

[44] *Ibid.* In *The Carthaginian Peace or the Economic Consequences of Mr. Keynes* (New York, 1952), Etienne Mantoux effectively demolishes some of Keyne's arguments on the political consequences of Versailles. But his rebuttal of Keyne's economic arguments is unconvincing.

[45] *Ibid.*, p. 48.

In actual fact, as everyone now knows, Germany was a net gainer of the financial transactions of the nineteen-twenties; she borrowed far more from private American investors (and failed to pay back) than she paid in reparations. This was of course little consolation to the German taxpayer, who was not at all the same person as the German borrower. For that matter, reparations gave little consolation to the taxpayers of allied countries, who immediately saw the proceeds transferred to the United States in repayments of war debts. Setting one thing against another, the only economic effect of reparations was to give employment to a large number of bookkeepers. But the economic facts about reparations were of little importance. Reparations counted as symbols. They created resentment, suspicion, and international hostility.[46]

The treaty of Vienna produced a century of relative peace. The treaty of Versailles produced only a truce of two decades; it "lacked moral validity from the start. It had to be enforced; it did not, as it were, enforce itself."[47] And the main actors of the new international system did not wish to enforce it. The Soviet Union was isolated behind the cordon sanitaire, the United States turned inwards and refused to join the League of Nations, Great Britain thought that it could return to its safe role as balancer. This left France and its client states against Germany and sundry minor revisionists. The geometry of power was unstable.

The end of World War I produced an overwhelming desire for a "return to normalcy." But the war had unleashed forces that were to preclude the return to the placid international system of the past century.

---

[46]Taylor, op. cit., p. 48.

[47]Ibid., p. 32.

Total war, the war by the masses, had transformed the <u>laissez-faire</u> state into the <u>welfare</u> state.

> Henceforth the functions of the nation-state were as much economic as political. The assumptions of these functions presupposed the abrogation of the international economic order and would, even if there had been no other obstacles, have prevented a revival of that order after 1919. Nationalism had invaded and conquered the economic domain from which the 19th century had so cunningly excluded it. The single world economy was replaced by the multiplicity of national economies, each concerned with the well-being of its own members.[48]

At first the main casualty of this new economic nationalism was the freedom of migration. "No single measure more clearly exhibited the inherent drive of the new and powerful labour interests towards policies of exclusive nationalism."[49] After the initial period of reconstruction the new international economic order seemed to function well. World trade expanded, unemployment was low, standards of living improved. And then came the Wall Street crash and the structure of international commerce collapsed. "The first international manifestation of the crisis was the complete cessation of American loans to Europe in the autumn of 1929; and this was rapidly followed by a drying up of purchasing power all over the world, resulting in a general catastrophic fall in prices."[50] Business cycles and depressions were no new occurence in international trade; according to theory the pure gold standard of the

---

[48]Carr, <u>Nationalism and After</u>, pp. 22-23.

[49]<u>Ibid.</u>, p. 27.

[50]E.H. Carr, <u>International Relations Between the Two World Wars, 1919-1939</u>. (London: Macmilland & Co., Ltd., 1965), p. 133.

nineteenth century was supposed to produce the necessary equilibrium among national economies at the expense of the least efficient. But this was the first depression of the new international system characterized by mass nationalism and by nation-states responsible to their mass electorates. National politics which had produced the welfare state disrupted the international economy. Most countries went off the gold standard.

> Moreover, these countries, in a desperate effort to keep their own agriculture and industry alive and to maintain a favourable balance of trade, were driven to every kind of expedient in the form of tariffs, import restrictions and quotas, export subsidies and exchange restrictions, amounting in some cases to a complete state control of foreign trade. The normal flow of commerce was almost completely interrupted. Unemployment figures rose everywhere by leaps and bounds. Half Europe was bankrupt, and the other half threatened with bankruptcy.[51]

Economic chaos and widespread unemployment produced social and political unrest. Communist parties with varying degrees of control from Moscow grew in strength. Most industrialized countries turned to the "left" but Germany, with a middle class that had been all but wiped out by the inflation of the early 1920's.(an inflation which "was more terrible that the French inflation of the eighteenth century")[52] turned to fascism. The major revisionist power acquired a totalitarian regime led by an individual who refused to play the rules of the international game,

---

[51] Ibid., pp. 133-134.

[52] Dehio, op. cit., p. 248.

even to the extent of using force in order to further foreign trade.

> In 1931, Great Britain established what came to be known as a "sterling bloc" by methods which were non-political and in appearance largely fortuitious. Germany in order to establish an equivalent "marc bloc" in Central and South-Eastern Europe, had to resort to methods which were frankly political and included the use and threatened use of force.[53]

With the rise of Hitler the international situation deteriorated rapidly. The fact that German rearmament was inadequate for a long World war, as witnessed by Germany's oil stocks and production, did not matter as long as others believed it was. But revisionism, mutual fear and armaments race were not the only source of instability; there was also conflicting legitimacy leading to ideological hostility.

> Before 1939 the international system was heterogeneous. A complex heterogeneity, moreover, since three regimes were in conflict, profoundly hostile to one another, each of them inclined to put its two adversaries "in the same sack." To the Communists, fascism and parliamentary governments were only two modes of capitalism. To the Western powers, communism and fascism represented two versions of totalitarianism. To the fascists, parliamentary government and communism, expressions of democratic and rationalistic thought, marked two stages in degeneration, that of plutocracy and that of despotic leveling.[54]

Before plunging into World War II, let us examine how the evolving international system affected the structure of the international oil industry. The end of World War I had shown to its participants

---

[53] Carr, The Twenty Year's Crisis, p. 129.

[54] Aron, op. cit., p. 115.

the vital importance of oil as a strategic commodity. All Great Powers
found it imperative to gain access to oil supplies but only the victors
were able to do so. Russia had nationalized all oil companies on its
soil and the United States was threatened with dwindling oil reserves.
The only known unexploited oil deposits lay in Mesopotamia. The dismemberment of the Ottoman Empire and the allocation, via mandate or sphere of
influence, of the Arabian Peninsula, and of Syria and Iraq to France
and Great Britain, preempted these deposits in favor of the latter. In
sharp contrast with the international law of the nineteenth century the
victors of World War I sanctioned the confiscation of private enemy
property, and the contractual rights of the Deutsche Bank in the Turkish
Petroleum Company were abolished and passed to the French Government.
As a result, Germany was effectively shut off from Middle East oil. It
was only the political, economic and predominant oil power of the United
States which was able to "open the door" of this government-sponsored
cartel.

The main political issue being settled, the international oil
industry developed in a framework of relative laissez-faire. This framework favored the economically strong and allowed the formation of restrictive agreements, such as the Red Line and Achnecarry agreements, without
undue government intervention. The age of motorization began. High
transportation costs kept energy competition to a minimum. As a result
tariffs had the main purpose of revenue raising. The United States continued to be the main oil exporter, the Middle East was directly or

indirectly under the control of the Great Powers, Indonesia the colony of the Netherlands. Venezuela and Mexico pursued their respective economic interests but could not maintain strong independent foreign policies. While the Soviet Union instituted bilateral commercial treaties, it behaved within the rules of the commercial game. The vast growth of the international oil industry was still ahead, but as a consequence of World War II it would never regain this relative political peace.

IV. THE PRESENT INTERNATIONAL SYSTEM

World War I began as a contest for limited aims among empires and gradually became a total war among nations and peoples. "In the second world war any valid or useful distinction between armed forces and civilian populations disappeared almost from the outset;. . . The individual had become little more . . . than a unit in the organized ranks of the nation."[55] Total mobilization of resources and populations, scorched earth policy saturation bombing, slave labor and genocide, all became official state policy. If Hitler did not succeed in his final attempt at Götterdämmerung, it was not for lack of trying. And in the Pacific war the kamikaze spirit was only squelched by the unleashing of the atom.

The Jewish population that had survived Hitler's extermination camps flocked to Palestine. Britain, the Mandate Power, had previously

---

[55]Carr, Nationalism and After, pp. 26-27.

followed the sensible policy of promising Palestine to both Jews and Arabs, but this balancing act could not be maintained in the face of mutually exclusive aims. The problem was dumped into the lap of the United Nations. In a rare show of unanimity (The Soviet Union wanted Britain out) the Great Powers agreed to the partitioning of Palestine. Israel was born and the following day began the war with the Arab League. The Balkans of the nineteenth century had been transplanted to the Middle East of the twentieth century.

At the end of World War II Europe and Japan lay in ruins. Into this power vacuum entered the two superpowers. For the first time since the Peloponnesian war the world had entered into a period of bipolarity. For the first time since the wars of religion, man's belief became a stake of conflict among states. Bipolarity and ideology -- man's secular religion -- created the phenomenon known as the Cold War. The scorpion and the tarantula, both armed with nuclear stingers, circled each other inside of the sealed bottle.

The superpowers organized the economic relations in their respective European spheres of influence. Having witnessed the social and political effects of the unparalleled destruction of Europe, the United States inaugurated the Marshall Plan which, in theory, was open to all European countries, including the Soviet Union. But the latter rejected it violently fearing that the aid would be used to challenge the Soviet hegemony in Eastern Europe. As a result American aid was channeled to

Western Europe alone; the only major condition being European economic cooperation via supranational organizations. The Soviet Union had been devastated by Germany. Consequently, instead of giving aid, it exacted reparations. East Germany paid directly; Eastern Europe indirectly, via mixed corporations, currency manipulations, administrative pricings, etc. More important, the Soviet Union imposed its model of industrialization and internal organization on its client states.

> Each satellite state was to resemble, economically, a microscopic Soviet Union, while the whole of Europe was to model itself on the United States. America's goals was to make its own aid unnecessary, thus restoring to the nations of Europe their lost independence (economic, not military); the Soviet policy tended to make definitive the economic dependence of the satellites, whose supply of raw materials could be assured only by the Soviet Union.[56]

The Marshall Plan chanelled some $20 billion into Europe. The years of reconstruction gave way to various "economic miracles." But some of the effects of World War II could not be undone. In rapid succession the overseas colonies peeled off from their respective mother countries. Britain took heed of the old master's advice that "prudent men make the best of circumstances in their actions, and, although constrained by necessity to a certain course, make it appear as if done by their own liberality."[57] But the French did not perceive the historic trend. It took the wars of Indochina and of Algeria to demonstrate to the French that their Fourth Republic had followed the footsteps of King Canute.

---

[56]Aron, op. cit., p. 456. Italics in the original.

[57]Niccolo Machiavelli, The Discourses, First Book, Chap. II.

It mattered little: by the early 1960's, and with the exception of
the remnants of Portugal's empire, all overseas colonies had received
the trappings of political independence, thus altering the structure of
the international system.

Having sketched some of the causes that led to the present
international system, let us analyze its main elements. According to
Hoffmann, these are bipolarity, an unprecedented heterogeneity in the
structure of the system, extreme variety in the domestic political systems,
sharp ideological clashes and a technological revolution. "At first
sight, the international milieu of the 1960s is almost a textbook example
of a revolutionary system."[58] We have already mentioned the novelty of
bipolarity; let us examine the effects of decolonialization.

> For the first time, the international system
> covers the whole planet, but it now includes ingre-
> dients of widely different origins and vintage.
> . . . Even though the basic unit of modern world
> politics is the nation-state, this over-all term
> conceals an extraordinary diversity. For one thing,
> modern nation-states differ in their degree of inte-
> gration. Some are states still in search of a nation,
> . . . others are national communities in a substantive
> sense as well. They also differ in their historical
> dimensions. . . Finally, they differ in size. . . .
> Because of this heterogeneity, the distribution of
> power between and around the United States and the
> Soviet Union is highly discontinuous and uneven.[59]

---

[58]Stanley Hoffmann, Gulliver's Troubles or the Setting of American Foreign Policy (New York: McGraw-Hill Paperbacks, 1968), p. 17.

[59]Ibid., pp. 17-18.

352.

Turning to domestic political systems, "The variety of regimes is so dazzling that political scientists cannot even begin to agree on classificatory schemes."[60] The international system is further racked by two types of ideological clashes: communism vs. the so-called free world, and anticolonial nationalism vs. the West. "These are doubly asymmetrical: in each case, one side is heavily ideological and the other rather pragmatic; in each case the ideological side is one the offensive."[61] And finally, "A technological revolution which contributes to instability both because it proceeds so fast and because it aggravates the unevenness among nations."[62]

With such disparate and destabilizing elements it is hardly surprising that the relations among the units of the present international system are immoderate in ends and means. But the key to the lack of system stability lies in its bipolar structure. "In a bipolar contest, prestige, power and security are rolled into one . . . When this contest is ideological as well, the distinction between a threat to the physical existence of the foe and a threat to his moral integrity becomes blurred."[63] By the same token, "A bipolar contest implies the risk of clashes at any point of contact between the two great powers and elsewhere in the search for supporters."[64] And insofar as these supporters are engaged in

---

[60] Ibid., p. 18.
[61] Ibid.
[62] Ibid., pp. 18-19.
[63] Ibid., p. 19.
[64] Ibid.

domestic conflicts they may embroil the superpowers in their respective quarrels. "The immoderation of means is due to the proliferation of conflict situations."[65]

The great paradox of the present international system is that while there has been an abundance of minor wars there has been no direct armed conflict among the superpowers. The key to this paradox lies in the revolution in military technology. Mutual second strike capability has achieved mutual deterrence and, therefore, a certain degree of stability in the pole of major tension of the international system. But by the same token mutual deterrence has injected the superpowers with a heavy dose of prudence which has weakened their diplomatic stance vis-à-vis medium and lesser powers.

> In short, the inhibition on a total use of coercive force has led to diversification and repression toward lower levels of violence. This development has also disconnected the once close link between mobilizable military potential and gains, even in the use of coercive power. Inversely, other ingredients of the supply of power -- geographical spread . . ., <u>the possession of scarce resources</u>, intangible assets of human talent . . . - are on their own perhaps for the first time, <u>in the sense of being fully exploitable even with no military force</u>.[66]

In his discussion of the concept of the international system Aron remarks that "only the actors performing in the plays belong to the troupe."[67] In the present international system the stars are so preoccupied

---

[65]Ibid.

[66]Ibid., p. 31. My emphasis.

[67]Aron, op. cit., p. 95.

with each other that the supporting cast has taken over the stage.
And the spectators are regaled with the antics of the starlets.

> As a result, a kind of de facto polycentrism occupies the forefront of the state, in which almost anyone who wants can play. . . . The 'centers' are states many of which lack the traditional ingredients of military might, but which are well supplied with the new factors of power and eager to play the game. . . . Because the new factors of power are complex and varied, their playing has a flexibility that seems to defy analysis.[68]

Having described the complexity of the present international system, let us analyze its influence on the structure of the international oil industry. Having established the systemic reasons for the present politization of all relations among states let us examine how the various elements of the international system affect the operating milieu of the industry. Bipolarity has as pervasive an influence on the international oil industry as it has on international relations. Both superpowers use the international oil (and gas) trade to maximize their respective current income and power. But their means are asymmetrical. The United States profits from the international oil trade via its control over American international oil companies that produce and sell foreign oil to foreign markets while the Soviet Union profits only from its oil exports. In view of their asymmetrical oil reserves and consumption the present situation clearly favors the United States. But as financial gains from this

---

[68]Hoffmann, Gulliver's Troubles, pp. 34-35.

middleman trade is dependent on large foreign investments, it produces a political liability that outweighs by far any political advantage in latent U.S. oil boycotts. As bipolarity implies competition in all countries the U.S. dominated international oil industry is vulnerable to Soviet efforts on many fronts.

The heterogeneity of the structure of the international system has had a profound effect on the structure of the international oil industry. For one thing the doubling of the number of independent states in two decades has vastly complicated the problems of the major international oil companies and of their respective home countries. Prior to World War II the international oil companies operated in Africa and Asia within the political framework of the various colonial or hegemonial powers. While this did not preclude commercial or political competition for supplies and markets, it did provide for political stability and for a consensus on the rules of the commercial game. Decolonialization has changed these rules. Every newly independent country, whether oil producer or consumer, must fit its participation in the international oil trade into its foreign economic and foreign policy objectives. And this complicates matters. The economic and political fragmentation of producing and consuming countries increases the bargaining power of the major oil companies but it also forces the duplication of production and consumption facilities to accommodate the respective national policies. And the multiplication of autonomous centers of decision requires a large degree of coordination from the international oil companies and from their respective home countries.

The heterogeneity of the structure of the international system is further compounded by the extreme variety in the domestic political systems and by ideological considerations. All states practice economic nationalism and use foreign economic policy in support of their respective general foreign policies. But in some states, mostly industrial, there is a consensus on the political framework that establishes the commercial rules, and on the general lines of foreign policy, while in the majority of the states there is a consensus on neither. When domestic changes, whether peaceful or violent, lead to changes in regimes (as opposed to government administrations) there is constant insecurity for domestic and foreign capital. When changes in regimes are further based on ideological clashes, they usually result in changes in foreign policy that lead to a reorientation in foreign trade and investment. And the oil industry is usually among the first to be affected.

In view of the nature of the subject the element of the international system that has most affected the international oil industry is the technological revolution. In regards to oil production, developments in oil exploration, drilling and production technology have had two effects. On one had they have allowed the increased finding and production of oil at lower cost which, together with improvements in tanker technology, have been largely instrumental in the displacement of coal from most of its markets. On the other hand these developments are a threat to the exports of the established c'l producers. Insofar as oil companies pursue their program of lessening of dependence on Middle East oil exporting countries, any technological development that allows them to discover new sources of oil or to produce more from existing "safe" sources, will assist them in these efforts. Present energy consumption technology has fully exploited the versatility of oil and transformed it into the leading source of energy. But technology does not stand still.

V. TOWARDS AN INORGANIC ENERGY SOCIETY

In Chapter III we saw that, provided the development of the required energy production and consumption technology, man can attain a virtual inorganic energy society in about half a century. Now let us see why he must strive to attain it sooner. There are two compelling reasons: ecology and conservation. Concern about the environment has suddenly become universal. Every form of pollution is now under attack, but perhaps the greatest criticism is reserved for the energy industries. Oil slicks produced by offshore drilling or by tanker wreckage pollute the seas and beaches, kill wildlife and disrupt the food chain of marine life. Nuclear stations produce thermal pollution, and may produce radiation levels harmful to man. Coal strip mining creates a blight on our landscapes and damages our agricultural potential. And the chimney stacks and exhaust pipes of our industrial civilization pollute our air. These scourges are bad enough as it is, but if we continue at our present rate of increase of air pollution we may alter our planet's carbon cycle.

> This is something more than a public health problem, more than a question of what goes into the lungs of an individual, more than a question of smog. The carbon cycle in nature is a self-adjusting mechanism. Carbon dioxide is, of course, indispensable for plants and is, therefore, a source of life, but there is a balance which is maintained by excess carbon being absorbed by the seas. The excess is now taxing this absorption and it can seriously disturb the heat balance of the earth because of what is known as the "greenhouse effect." A greenhouse lets in the sun's rays but retains the heat. Carbon dioxide, as a transparent diffusion, does likewise. It keeps the heat at the surface of the earth and in excess modifies the climate. It has been estimated that, at the present rate of increase (6 billion tons a year), the mean annual temperature all over the world might increase by 3.6 degrees centigrade in the next forty to fifty years.[69]

"During the past century . . . more than 400 billion tons of carbon have been artificially introduced into the atmosphere."[70] And the effects are beginning to tell. "The northpolar ice-cap is thining and shrinking. The seas, with their blanket of carbon dioxide, are changing their temperature, with the result that marine plant life is increasing and is transpiring more carbon dioxide."[71] If out of stupidity or greed we persist in our present course, we will reap climatic changes of uncalculable results. "One would be well-advised not to take

---

[69]Lord Ritchie-Calder, "Mortgaging the Old Homestead," Foreign Affairs, Vol. 48, No. 2, January 1970, pp. 214-215.

[70]Ibid., p. 214.

[71]Ibid., p. 215.

ninety-nine year leases on properties at present sea levels."72

Let us assume that public concern will result in legislation that will override vested interests. Devices that stop the emission of carbons will be installed at the expense of the consumer; chimney stacks and exhaust pipes will cease polluting the air. Do we have the right to rest on our laurels and to leave the outcome of our energy production and consumption pattern to the interplay of market forces with only a "normal" level of government intervention?

It is all a matter of perspective. On one hand our planet's fossil reserves amount to untold billions of tons that can last mankind for centuries even at the present rate of increase in consumption. On the other hand the reckless use of our geological capital will deprive the descendants of our species of precious non-renewable natural resource. "As long ago as 1872, the famous Russian chemist Mendeleyev, after a visit to the Pennsylvania oilfields, told his government that: 'This material (oil) is far too precious to be burnt. When burning oil we burn money; it should be used as a chemical base material.'"73 The petro- and carbochemical industries are only beginning to explore their potentials. And our planet could sustain life for another five billion years until our sun goes nova . . .

72Ibid. There is an even more chilling possibility. When particulate matter, or dust -- a byproduct of industry and of burning of coal and oil -- is introduced into the atmosphere, it "reflects light from the sun back into space and tends to cool the climate." (The New York Times, April 30, 1970). As a result of this effect some scientists are predicting the advent of a new ice age. While scientists are not sure which of the

By increasing the cost of our energy use we can drastically reduce the pollution of our air. But short of a decrease in energy consumption the only thing that we can do within our present technological level is to increase energy production and consumption efficiency. What we must do is to provide the necessary funds for accelerated and intensified research that will allow us to attain a new technological plateau. And once attained we will then be in a position to follow the categorical imperative of wise management of our fossil resources. Only the fusion of the atom will satisfy mankind's needs for cheap and virtually unlimited energy. And the fusion torch will allow us the wise management of our ores resources as well.

> To maintain and expand our high-level society, man is rapidly exhausting his geological capital which has been accumulated over hundreds of millions of years in the form of ores and fuels. The use of the fusion torch is not only an alternative to pollution problems, but also has the potential, using energy from controlled nuclear fusion, for a closed-cycle economy which can solve the materials problems of the world by simply circulating a material from one form to another.[74]

---

two, floods or ice age, will occur first, they do not think that both effects will cancel each other out.

[73] The Petroleum Handbook, compiled and published by Shell, cited in Tugendhat, op. cit., p. 120.

[74] Eastlund and Cough, op. cit., p. 17.

But pending the development of fusion reactors and the transformation of all our major thermal and mechanical energy consuming equipments to the use of electricity, we will have to satisfy a large part of our ever increasing demands for energy from organic sources. At our present technological level we have only a limited range of options. Having for the present no realistic alternatives, we can then switch our attention to the next four or five decades. From this perspective our organic energy reserves are more than abundant. How will the international society of the coming decades organize, in order to satisfy its energy demands? The answer lies in the realm of prophecy. But one may venture some speculations.

While there is only a limited range of options in regards to the aggregate world demand for organic energy sources the role of oil in international relations might be significantly altered by a change in the pattern of international oil trade. Such a change might come about by an increased diversification of oil exporting countries and/or by a successful conversion of fossils to oil and gas. Let us explore some of the possibilities and their likely effects on the international oil trade of the coming decades.

We may begin by spinning the geological roulette. A successful run against the casino bank requires a willingness to play and some means of improving the gamblers' odds. Taking the gamblers' compulsion for granted let us examine some of their possible tricks. While the oil industry has not given hope on the eventual discovery of a divining rod it continues its pursuit in the improvement of exploration technology. This pursuit is unlikely to bring any single dramatic result but rather a cumulative effort leading to an enhanced exploration capability. A similar forecast can be made for improvements in production technology. On the other hand, improvements in offshore drilling, completion and production technology, and in oil and gas transportation technology from polar areas will pave the way for oil production from new and potentially significant areas.

But regardless of technological improvements oil will still be where you find it. And while the roulette has neither memory nor premonition one can venture to estimate the reach of the croupier's rake. The pattern of the international oil trade of the coming decades will be influenced by the location, quantity, production costs and chemical composition of future oil discoveries. While these factors have obvious economic implications the factor of location has an additional political dimension insofar as it involves diversification of sources of supply. Thus, while a major new oil find in Algeria would result mainly in an accrual of reserves with only marginal export increase, the same find in neighboring Tunisia would probably result in the formation of a new export center at the expense of some production increase of traditional oil exporters. This, in turn, would increase the bargaining power of the importers and presumably of the major oil companies vis-à-vis OPEC members, cause additional internal conflict resolution problems for OPEC, and further decrease the effectiveness of oil exporters' boycotts.

A similar analysis may be made for offshore production within national jurisdiction. One could thus contrast the limited impact of further offshore production in the Persian Gulf to a major trade reorientation involved in a possible major oil strike in the North Sea. But what about production beyond national jurisdiction? In the absence of an international agreement such production would result in economic benefits for the few private and/or public companies with the necessary capital and technological capability, and to their respective home countries, and in the further dillution of economic and political power of traditional oil exporters. A similar effect might be achieved in the case of an international agreement along the lines of the Malta proposal, where an international licensing authority would receive a royalty that would be syphoned off to a United Nations development fund. Such an agreement would put a floor on the price of offshore oil beyond national jurisdiction but leave the bargaining power of the major companies intact. But what if an international agreement created an authority that could regulate price, volume and distribution, and hamper the operational freedom of the international majors in that area? The international economic and political possibilities are staggering, especially if one assumes a relatively large world oil production from offshore areas beyond national jurisdiction controlled by an international authority based on the one country-one vote principle.

364.

The impact of a successful conversion technology of fossils to oil and gas on the international oil trade of the coming decades is difficult to gauge. A series of technological breakthroughs leading to dramatic cost reductions in the conversion of all fossils to oil and gas could, when combined, isolate the major world oil importers from any significant participation in international oil trade. But quite apart from technological difficulties such an outcome is unlikely in view of the transportation costs from the oil shale field in Colorado, or from the Athabasca tar sand fields, to the major markets, and in view of the likely competition from liquid oil. In other words these fossils should prove to become complements to oil, but hardly a substitute.

Having examined some of the possibilities for the international oil trade of the coming decades we may well inquire whether there is a preferred outcome. But first some assumptions. Barring unlikely outcomes such as nuclear war among the superpowers, a return to the cold war leading to a renewed melting of sovereignty among blocs, a duopoly that would maintain world order on the basis of overwhelming force, or a vast proliferation of nuclear powers, the international system for the period under consideration will continue to be <u>revolutionary</u>. One could expect an increased moderation in the relations among the superpowers, a certain amount of regional cooperation, perhaps even some form of multi-polar system; but "it is all too easy to imagine a multi-hierarchical system of dizzying instability."[75] In this likely revolutionary and unstable

---

[75] Hoffmann, <u>Gulliver's Troubles</u>, p. 357.

international system with a largely unusable amount of military power, relations among states will be immoderate in both ends and means. Consequently, foreign economic policy will continue to be a major tool of foreign policy in general. "In a world that must solve internationally the problems once tackled either by domestic or private transnational channels, there will be different hierarchies for different tasks, corresponding to different computations of power."[76] And the power to withhold oil supplies will certainly be part of the functional hierarchical structure.

The nature of the system determines the moral standards for its judgment. In a system composed of units with different standards of legitimacy using most of the means at their disposal in order to further their conflicting ends, all means become politicized and inseperable from the ends. An impartial observer can only examine the forces at work in the international milieu and having determined the systemic causes that produce conflicting ends among its units, reserve any judgment on the means. Neither can he take a stand on any arbitrary arrangements among conflicting interests such as prevalent in the international oil industry without imposing his standards of legitimacy on others. The ideals of this observer are transnational. His feelings are for individuals, not for collectivities.

---

[76]Ibid., p. 356.

Having stated the assumptions for the subsequent analysis, let us examine whether there is any preference in the way the international society of the following decades will satisfy its energy needs. From the point of view of the international system is there any preference for the future political economy of international oil? Let us begin with economics, with the distribution of profits from the international oil trade.

OPEC member countries want more. They justify their claim on the basis of their sovereignty, including their permanent sovereignty over their natural resources, on their need for money for their economic development, and on the fact that oil is a limited and exhaustible resources. The basis of their claim is legitimate because it is " . . . a universally recognized principle of public law and has been repeatedly reaffirmed by the General Assembly of the United Nations, most notably in its Resolution 2158 of November 26, 1966."[77] In the operative paragraph of this resolution, the General Assembly:

> <u>Recognizes</u> the right of all countries, and in particular of the developing countries, to secure and increase their share in the administration of enterprises which are fully or partly operated by foreign capital and to have a greater share in the advantages and profits derived therefrom on an equitable basis, with due regard to the development needs and objectives of the peoples concerned and to mutually contractual practices, and calls upon the countries from which such capital originates to refrain from any action which could hinder the exercise of that right.[78]

---

[77]Preamble of Resolution XVI.90, <u>Declatory Statement of Petroleum Policy in Member Countries</u>, at the OPEC's 16th Conference, Vienna, June 24-25, 1968.

367.

In other words, the developing countries, who constitute the majority of the General Assembly, fully agreed that they are entitled to more. Undeterred by their fiasco at the 1964 UNCTAD Conference, they persisted in their attempts to parlay their voting majority into cash. This is quite understandable, but the mere fact that these attempts are successfully resisted deprives this resolution of universal legitimacy.

The General Assembly grants the developing countries the right to a greater share. OPEC improves on the theme by granting its member governments the right to <u>determine the price</u>. In its Resolution XVI.90, <u>Declaratory Statement of Petroleum Policy in Member Countries</u>, OPEC recommends:

> All contracts shall require that the assessment of the operator's income, and its taxes or any other payments to the State, be based on a posted or tax reference price for the hydrocarbons produced under the contract. <u>Such price shall be determined by the Government</u> and shall move in such a manner as to prevent any deterioration in its relationship to the prices of manufactureed goods traded internationally. . .[79]

We have repeatedly noted that price fixing of oil by producer governments is ineffective unless accompanied by a successful cartel, and that the realization and maintenance of such a cartel is a chimera because of the conflicting interests of its would-be members. We need, therefore, comment no further on OPEC's desires in that respect. But price fixing

---

[78]General Assembly Resolution 2158 (XXI) on the Report of the Second Committee (A/6158) - of 25 November 1966: <u>Permanent Sovereignty over National Resources</u>, I, No. 5.

[79]Resolution XVI.90, <u>op. cit.</u>, "Posted Prices or Tax Reference Prices," my emphasis.

of oil at an artificially high level could be feasible as part of a commodity agreement including producer and consumer governments. Would it be desirable?

> If commodity control is to be used in this way to stabilize prices at an artificially high level, the results will be disastrous to the world's economy and to the growth of the world's wealth at the maximum rate. It is a fundamental economic principle that prices should be related to cost of production. If they are not, resources are being wasted somewhere. Demand, supply and price are always altering somewhere, and cost of production is the only true measuring rod by which the world's resources can be continually adjusted to the world's needs.[80]

Granted that such a commodity control divorced from considerations of production costs would be injurious to the world economy as a whole; a case may still be made to favor developing countries in the name of social justice. Most economists, unless they are of the "inherent value" (whether of labour, oil or peanuts) school of thought, would agree that commodity trade " . . . at high prices divorced from any relationship with costs or production - prices as high as the richer nations can be coerced into paying by political and any other pressures- . . . is really only charity in disguise."[81] But even those opposed to foreign aid will agree that the Robin Hood image has popular appeal. A suburban commuter stalled in traffic may take some comfort from the knowledge that the increased price for the gasoline he is needlessly burning in an inefficient and air-polluting engine is being used to pay for fish flour for a

---

[80]Rowe, op. cit., p. 215.

[81]Ibid.

protein-starved infant in some nomad's tent. But is it?

The question is fundamental. Is foreign aid under the guise of foreign trade legitimate? Needless to say the answer depends on standards of legitimacy. Let us assume equal transfer of resources through either channel. In the case of conventional foreign aid, whether bilateral or multilateral, resources are tied to some purpose that both donor and receiver conceive as legitimate. If foreign aid is channeled through foreign trade, the legitimacy of the transaction depends on the use of the resources by the receiver. And this requires judgment in the domestic and foreign policies of other nations, a rather thankless task in an unstable, revolutionary, international system where actors agree on neither ends nor means.

But let us try. According to both the United Nations General Assembly and OPEC the purpose of increased receipts from commodity trade is economic development. Despite a cascading literature on the subject of development, there is little consensus on what it is or how to go about it. Nevertheless, there are some economic indicators that, together with less empirical values, give an idea of a country's economic progress. Despite shortcomings, and taking a normal amount of inefficiency and corruption into consideration, Venezuela, Iran and Indonesia (since the ouster of Sukarno) have shown economic progress. So did Libya under King Idris. But his successors have decided to invest a large part of Libya's oil receipts in Mirage jets. And regardless of its qualities and virtuosity Dassault's brainchild is not digestible.

This line of argument is leading to a <u>reductio</u> <u>ad</u> <u>absurdum</u>. As every United Nations delegate, with the possible exception of Israel's, would retort, all countries are sovereign to spend their money as they please. And so they are. Any judgement in this respect requires a standard of legitimacy and taking a position on the relative merits of any nation's accrual of power. But the insistence on more money from trade for the purpose of economic development and the declaration of sovereignty when the money is diverted to other purposes tests the canons of logic. Granted that nations possess the sovereign right of contradiction, oil-consuming nations still have the sovereign right to veto any commodity agreement that would result in foreign aid for purposes not to their liking.

In fact, any <u>consistent</u> tying of foreign trade receipts with economic development would result in chaos. Assuming agreement on some standards of economic development, one would then require that international oil trade be conducted according to some means test for both oil exporting and importing nations. We may veil the curtain of charity on this political, economic and administrative nightmare. But we may conclude that in view of the absence of concrete agreement on common purposes, the distribution of profits from the international oil trade can be left to the interplay of geology, economics and politics.

We now turn our attention to the international oil companies. "What are the essential - and irreducible - functions of this prodigious middleman, the international oil industry? And how much return does one need to offer capital to enter or stay in it?"[82] According to Hartshorn:

> . . . the essential operational function of an internationally integrated oil company may seem to be the logistic - the disposition of supplies from many sources to meet many different demands, and the setting of relative prices. Its main technical and social function may be one of selecting and training good managers of all nationalities and imbuing them with wider than national experience in management. Its main financial function is perhaps that of acting as a specialized investment bank with exceptional experience in a particularly risky and competitive technology, with a long enough view to ensure the steady development of an internationally essential resources. . .[83]

How much are these functions worth to the world economy? The answer is unquantifiable but if one were to take efficiency as a criterion of world benefit one could hardly imagine governments performing a more efficient task. In fact the question is pointless. Considering that the price of world oil is only vaguely related to production costs, but is rather the product of often conflicting political pressures, a more relevant question would be what is the function of the international oil companies in the pricing mechanism and in the allocation of benefits from the international oil trade? According to Hartshorn, and I agree, the function is that of a buffer among conflicting interests. "It can

---

[82]Hartshorn, op. cit., p. 376.

[83]Ibid., pp. 377-78.

act as a cushion, if not an arbiter, between the claims of different countries, without the political friction that continous confrontation of national interests would cause."[84]

The main theoretical objection to the buffer function of the international oil companies is that it rests on the assumption that the avoidance of political friction and of continuous confrontation is in the interests of all. This assumption is untenable. The international oil companies may well be a buffer among conflicting interests but they are also economic powers in their own right who pursue their goals of maximization of international integrated profits and of preservation, if not increase, of their respective market shares. Consequently, their buffer function favors the <u>status quo</u> at the expense of the revisionists. "In practice, however, international firms must bow to the wishes of those governments that are in the strongest position to bring pressures on them."[85]

The international oil companies perform a constant balancing act among the interests of shareholders, employees, consumers and the general public, and among producer, consumer and home countries. To the extent that they perform their function of providing a large part of the world with an indispensible commodity their function may be deemed as legitimate. But this does not mean that the balance of power among the various interests should remain static in the name of

---

[84]<u>Ibid.</u>, p. 387.

[85]Penrose, <u>op. cit.</u>, p. 268.

stability or of efficiency. And this brings us to the political aspects.

Concern about the world economy is similar to concern about world peace: something better left for others. Economic theory teaches about the benefits derived from international trade, and the world depression is a stark reminder of what can happen when the machinery breaks down. But the gap between international <u>laissez-faire</u> and autarky is large enough to allow foreign trade to be used in support of power relations. In an international system where the international flow of capital and goods is accompanied by the flow of power, the concept of efficiency of allocation of resources loses much of its theoretical, or even normative, significance: resources can be measured, power can only be inferred. Consequently the use of efficiency of allocation of world resources as a criterion of universal desirability resembles the use of the concept of stability in international relations. It not only favors one nation at the expense of another but inevitably impinges on some power relations. Efficiency may benefit all nations in the long run but nations and their politicans have the sovereign right to choose between their long-term and short-term interests. Seen from this perspective it matters little whether the oil trade of the following decades will take place on a bilateral, multilateral or most-favored-nation basis, or on a combination of the above. Every theoretical permutation has its economic and political advantages: arbitrary price balanced by intangible influence. The further spread of bilateral oil trade may complicate logistics, restrict the operational freedom of the international oil companies, and even deprive them of profits, but this is a small price to pay for the thrill of declaiming on the stage, and even more so when the price is paid by others.

Would it be desirable to take international oil trade "out of politics?" In an unstable, revolutionary international system the question is either naive or slanted towards the interests of those who stand to lose most from political pressures. And in view of the many types of possible political pressures, intractable as well; "the animals at Madariaga's imaginary conference found that 'the lion wanted to eliminate all weapons but claws and jaws, the eagle all but talons and beaks, the bear all but an embracing hug.'"[86] Asking the Middle East Arab oil importers and transit countries to refrain from the threat of oil boycotts in the name of stability is asking them to deprive themselves of their only weapon and of their main source of influence in international affairs. But by the same token one can neither deprive the international oil companies of their weapon of diversification of sources of supply, especially if it results in the spreading of benefits from oil exports to other developing nations. Nor can one ask the oil importing nations to leave their industry and transport at the mercy of the foreign policies of oil exporting nations and of foreign oil companies. Finally, asking the U.S. to relinquish control over the foreign operations of the

---

[86] quoted in Inis L. Claude, Jr., <u>Swords into Plowshares, The Problems and Progress of International Organization</u>, third edition (New York: Random House, 1964), p. 271.

American international majors is to ask it to relinquish the use of its economic power as a tool of its foreign policy. And so the game goes on, amoral because of the absence of harmony in anarchy.

Let the decades unfold and energy competition take its course. Spurred by concerns over pollution and ecology, and by economic and political considerations, the international society will unlock the energy derived from the fusion of the atom and develop the means to satisfy most of its energy demands from electricity. The industrial countries will be the first to spearhead the movement towards the abandonment of fossils as a source of energy. The combination of vast energy demand, balance of payments drain, political dependence on energy exporters, and superior technological capability, will act as a spur to attain the goal of virtual energy autarky. And as the energy production and consumption technologies are established they will be transplanted to the less industrialized countries. It may be hoped, evn if doubted, that the developing countries who depend on organic energy exports for their development will have attained a degree of industrialization and of diversification of trade that will cushion the impact of the loss of their major export commodity. But "progress does not and cannot mean equal and simultaneous progress for all."[87] As for the oil companies, they will have ample time to transform themselves into combined energy and chemical companies.

---

[87] Edward Hallet Carr, What is History (New York: Alfred E. Knopf, 1962), p. 155.

The political economy of the international energy trade of the future inorganic energy society will be vastly simplified. Apart from some moderate amounts of organic energy for minor markets unconnected to power grids, and for specialized uses, energy trade will consist of uranium and thorium, perhaps of plutonium, lithium and tritium, definitely of electric power across border areas, but hardly of water, the source for deuterium. One can visualize some chicanery in licensing and patenting, political strings attached to loans for fusion reactors and spare parts, etc. But the fuel that fires today's political passions will be spent. Future history students will compare the trade in oil with the trade in spices, pipelines with the silk route, and refineries with the oven of the alchemist.

The technological level of energy production and consumption defines a level of civilization and may give a strong indication of historical progress. But how does it relate to freedom, social justice and peace? In contrasting our ancestors' discovery of the production of fire by friction with the steam engine, Engels wrote about a century ago that ". . . all past history can be characterized as the history of the epoch from the practical discovery of the transformation of mechanical motion into heat up to that of the transformation of heat into mechanical motion."[88] Some future political philosopher, disgruntled by the state of his society, might equally dismiss the era of the liquid hydrocarbon

---

[88]Friedrich Engels, <u>Herr Eugen Dühring's Revolution in Sciences</u>, Chapter XI.

as a mere bridge between the combustion of the carbon and the fusion of the hydrogen. And he may well have grounds for hope. Future energy technologies using matter-antimatter reactions or even some form of portable matter-energy transmuters might well produce cornucopias to satisfy all production needs. But what about their distribution? As long as man, singular or collective, measures himself in relation to others, there will always be a relative scarcity of material and psychological values leading to discontent about his lot. Some reactions to this relative scarcity may perhaps be abated by the heralded biological revolution. But from our perspective we can conclude that man will reap the advantages of his applications of the theory of relativity within the boundaries set by his applications of the theory of relative advantage.

SELECTED BIBLIOGRAPHY

This bibliography consists of three parts: (a) <u>Oil Technology, Economics and Politics</u>; (b) <u>Energy</u>; (c) and other social science works broadly classified under the term <u>History and Politics</u>. This classification is a compromise between various possible solutions, each unsatisfactory for a different reason, such as classification by chapter, topic, discipline, etc. All secondary references, as well as references to newspapers and periodicals, have been deleted.

OIL TECHNOLOGY, ECONOMICS AND POLITICS

Adelman, Morris A. "The World Oil Outlook," <u>Natural Resources and International Development</u>, Marion Clawson, editor. Essays from the Fifth Annual Resources for the Future Forum. Baltimore: The Johns Hopkins University Press, 1965.

_____. "Oil Prices in the Long Run (1963-75)," <u>Journal of Business of the University of Chicago</u>, April 1964.

The Chase Manhattan Bank, N.A. <u>Balance of Payments of the Petroleum Industry</u>. New York: The Chase Manhattan Bank, 1964.

Clark, James A. <u>The Chronological History of the Petroleum and Natural Gas Industries</u>. Houston: Clark Book Co., 1963.

Flandrin, Jacques and Chapelle, Jean. <u>Le Pétrole</u>. Paris: Technip, 1961.

Fontaine, Pierre. <u>La Guerre Oculte du Pétrole</u>. Paris: Dervy, 1949.

Foursenko, A. A. <u>Neftianie Tresti i Mirovaia Politika. 1880-e godi - 1918g</u>. Moscow: Nauka, 1965.

Frank, Helmut, J. <u>Crude Oil Prices in the Middle East. A Study in Oligopolistic Price Behavior</u>. Praeger Special Studies in International Economics and Development. New York: Praeger, 1966.

Frankel, P. H. <u>Essentials of Petroleum</u>. London: Chapman and Hall, 1940.

_____. <u>Mattei: Oil and Power Politics</u>. New York: Praeger, 1966.

_____. <u>Oil: The Facts of Life</u>. London: Weidenfeld & Nicolson, 1962.

_____. "The Relation of World Oil to Developments in the USA", Testimony submitted to the Subcommittee on Antitrust and Monopoly of the U.S. Senate, March 26, 1969.

Gerretson, Frederick Carl. *History of the Royal Dutch*. Leiden: E.J. Brill, 1953.

Gupta, Raj Narain. *Oil in the Modern World*. Allahabad: Kitab Mahal, 1949.

Hartshorn, J.E. *Politics and World Oil Economics. An Account of the International Oil Industry in its Political Environment*. Revised Edition. New York: Praeger, 1967.

Lenczowski, George. *Oil and State in the Middle East*. Ithaca, New York: Cornell University Press, 1960.

Lieuwen, Edwin. *Petroleum in Venezuela. A History*. New York: Russell & Russell, 1967.

Lubell, Harold. *Middle East Oil Crises and Western Europe's Energy Supplies*. A Rand Corporation Research Study. Baltimore: The Johns Hopkins University Press, 1963.

Lufti, Ashraf. *OPEC Oil*. Middle East Oil Monographs: No. 6. Beirut: Middle East Research & Publishing Center, 1968.

National Petroleum Council. *Impact of New Technology on the U.S. Petroleum Industry, 1946-1965*. Washington: National Petroleum Council, 1967.

_____. *Petroleum Resources Under the Ocean Floor*. Washington: National Petroleum Council, 1969.

Odell, Peter R. *An Economic Geography of Oil*. Praeger Surveys in Economic Geography. New York: Praeger, 1963.

Organization for Economic Cooperation and Development. *Oil Today (1964)*. Paris: O.E.C.D., 1964.

_____. *Pipelines and Tankers. A Report on the Effect of the Use of Pipelines on the Transport of Oil by Tankers*. Paris: O.E.C.D., 1961.

Organization of the Petroleum Exporting Countries. *Review and Record*. Years 1965-1969. Vienna: O.P.E.C.

Parra, Francisco R. *The Development of Petroleum Resources Under the Concession System in Non-Industrialized Countries*. OPEC Pamphlet EC/64/III. Geneva: O.P.E.C., 1964.

Penrose, Edith T. *The Large International Firm in Developing Countries. The International Petroleum Industry*. London: George Allen and Unwin Ltd., 1968.

Rowe, J.W.F. *Primary Commodities in International Trade.* Cambridge: University Press, 1965.

Schwadran, Benjamin. *The Middle East, Oil and the Great Powers.* New York: Praeger, 1955.

Tanzer, Michael. *The Political Economy of International Oil and the Underdeveloped Countries.* Boston: Beacon Press, 1969.

Tugendhat, Christopher. *Oil: The Biggest Business.* New York: G.P. Putnam's Sons, 1968.

United Nations Conference on Trade and Development. Second Session. New Delhi. *Commodity Problems and Policies.* TD/97, Vol. II. New York: United Nations, 1968.

United Nations Department of Economic and Social Affairs. *Utilization of Oil Shale. Progress and Prospects.* ST/ECA/101. New York: United Nations, 1967.

United Nations General Assembly. *Report of the Ad Hoc Committee to Study the Peaceful Uses of the Sea-Bed and the Ocean Floor Beyond the Limits of National Jurisdiction.* General Assembly Official Records Twenty-Third Session. A/7230. New York: United Nations, 1968.

United States Department of the Interior. *United States Petroleum Through 1980.* Washington, D.C.: U.S. Government Printing Office, 1968.

_____. Geological Survey Circular 522. Hendricks, T.A., *Resources of Oil, Gas, and Natural-Gas Liquids in the United States and the World.* Washington, D.C.: U.S. Government Printing Office, 1965.

_____. Geological Survey Circular 523. Duncan, Donald C. and Swanson, Vernon E. *Organic-Rich Shale of the United States and World Land Areas.* Washington, D.C.: U.S. Government Printing Office, 1965.

ENERGY

Bishop, Amasa S. "Recent World Developments in Controlled Fusion." Paper read at the Plasma Physics Division of the American Physical Society, November 12, 1969.

_____. "The Status and Outlook of the World Program in Controlled Fusion Research." Paper read at the National Research Council of the National Academy of Sciences, November 11, 1969.

_____. "Conference Summary." Paper read at the Symposium on Nuclear Fusion Reactors at Culham, England, September 26, 1969.

The Chase Manhattan Bank. *Outlook for Energy in the United States.* New York: The Chase Manhattan Bank, 1968.

Eastlund, Bernard J. and Gough, William C. *The Fusion Torch. Closing the Cycle from Use to Reuse.* U.S. Atomic Energy Commission, Division of Research. Washington, D.C.: U.S. Government Printing Office, 1969.

European Economic Commission. *Première Orientation pour une Politique Énérgétique Communautaire.* (Communication de la Commission au Conseil). COM (68) 1040. Brussels: E.E.C., December 18, 1968, and Annexes I and II, January 17, 1969.

_____. *Tendances Énérgétiques Mondiales.* Serie Énergie No. 1. Brussels: E.E.C., 1968.

Kohler, Foy D. and Harvey, Mose L. "On Appraising Soviet Science and Technology," *Interplay. The Magazine of International Affairs.* Vol. 3 No 4, November 1969.

Malvestiti, Piero. *Sources of Energy and Industrial Revolutions.* First published by the Bocconi University, Milan, reprinted by the European Community Information Service.

Manners, Gerald. *The Geography of Energy.* London: Hutchinson University Library, 1964.

National Coal Association. *Bituminous Coal Facts 1968.* Washington, D.C.: National Coal Association, 1969.

Organization for Economic Cooperation and Development. *Energy Policy. Problems and Objectives.* Paris: O.E.C.D., 1966.

Ritchie-Calder, Lord. "Mortgaging the Old Homestead," *Foreign Affairs,* Vol. 48, No. 2, January 1970.

Rose, D.J. *Engineering Feasibility of Controlled Fusion.* MIT-3980-6, Nuclear Fusion 9 (1969). Cambridge, Mass.: M.I.T., 1969.

Seaborg, Glenn T. "Fission and Fusion - Developments and Prospects." Paper read at the Council for the Advancement of Science Writing, Berkeley, California, November 20, 1969.

Schurr, Sam H. and Netschert, Bruce C., et al. Energy in the American Economy, 1850-1975. Its History and Prospects. Baltimore: The Johns Hopkins University Press, 1960.

United States Atomic Energy Commission, Division of Technical Information, Series "Understanding the Atom." Oak Ridge, Tenn.: U.S.A.E.C.

_____. Corliss, William R. Direct Conversion of Energy. 1967

_____. _____. SNAP: Nuclear Space Reactors. 1966.

_____. Donnely, Warren H. Nuclear Power and Merchant Shipping, 1965.

_____. Glasstone, Samuel. Controlled Nuclear Fusion. 1968.

_____. Hogerton, John F. Atomic Fuel, 1964; and Nuclear Reactors, 1967.

_____. Lyerly, Ray L. and Mitchell, Walter III. Nuclear Power Plants, 1968.

_____. Murrows, Grace. Nuclear Energy for Desalting, 1967.

World Power Conference. World Power Survey of Energy Reserves, 1968. London: World Power Conference, 1969.

HISTORY AND POLITICS

Aron, Raymond. Peace and War. A Theory of International Relations. Richard Howard and Annette Baker Fox, translators. New York: Doubleday & Company, Inc., 1966.

Carr, Edward Hallett. International Relations Between the Two World Wars, 1919-1939. London: Macmillan & Co., Ltd., 1965.

_____. Nationalism and After. London: Macmillan & Co., Ltd., 1945.

_____. The Twenty Years' Crisis, 1919-1939. An Introduction to the Study of International Relations. New York: Harper & Row, 1964.

_____. What is History? New York: Alfred A. Knopf, 1962.

Cohen, Benjamin J., Editor. American Foreign Economic Policy. Essays and Comments. New York: Harper & Row, 1968.

Dehio, Ludwig. *The Precarious Balance. Four Centuries of the European Power Struggle*, Charles Fullman, translator. New York: Random House, 1962.

Engels, Friedrich. *Herr Eugen Dühring's Revolution in Science*. Moscow: Foreign Languages Publishing House, 1954.

Hoffmann, Stanley H. *Contemporary Theory in International Relations*. Englewood Cliffs, New Jersey: Prentice-Hall, Inc., 1960.

_____. *Gulliver's Troubles, Or the Setting of American Foreign Policy*. A volume in the series "The Atlantic Policy Studies" Published for the Council on Foreign Relations. New York: McGraw-Hill Book Co., 1968.

_____. *The State of War. Essays on the Theory and Practice of International Politics*. New York: Praeger, 1965.

Keynes, John Maynard, *Economic Consequences of the Peace*. New York: Harcourt, 1920.

Kirk, George E. *A Short History of the Middle East From the Rise of Islam to Modern Times*. Seventh Revised Edition. New York: Praeger, 1964.

Lenin, V.I. *Imperialism, the Highest Stage of Capitalism. A Popular Outline*. Peking: Foreign Language Press, 1965.

Morgenthau, Hans J. *Politics Among Nations. The Struggle for Power and Peace*. Fourth Edition. New York: Alfred E. Knopf, 1967.

Nkrumah, Kwame. *Neo-Colonialism, the Last Stage of Imperialism*. New York: International Publishers, 1965.

Schumpeter, Joseph A. *Capitalism, Socialism and Democracy*. Third Edition. Harper Torchbooks. New York: Harper & Row, 1962.

Taylor, A.J.P. *The Origins of the Second World War*. Second Edition. New York: Fawcett World Library, 1966.

Waltz, Kenneth N. *Man, The State and War. A Theoretical Analysis*. New York: Columbia University Press, 1965.

ENERGY IN THE AMERICAN ECONOMY

An Arno Press Collection

Abdallah, Hussein. **The Market Structure of International Oil With Special Reference to the Organization of Petroleum Exporting Countries.** (Doctoral Thesis, University of Wisconsin, 1966). 1979

Allain, Louis John Joseph. **Capital Investment Models of the Oil and Gas Industry.** (Doctoral Thesis, Purdue University, 1970). 1979

Bakerman, Theodore. **Anthracite Coal: A Study in Advanced Industrial Decline.** (Doctoral Dissertation, University of Pennsylvania, 1956). 1979

Barnett, Harold J. **Atomic Energy in the United States Economy.** (Doctoral Thesis, Harvard University, 1952). 1979

Becker, Clarence Frederick. **Solar Radiation Availability on Surfaces in the United States as Affected by Season, Orientation, Latitude, Altitude and Cloudiness.** (Doctoral Thesis, Michigan State University, 1956). 1979

Bentley, Jerome Thomas. **The Effects of Standard Oil's Vertical Integration into Transportation on the Structure and Performance of the American Petroleum Industry, 1872-1884.** (Doctoral Dissertation, University of Pittsburgh, 1974). 1979

Bouhabib, Abdallah Rashid. **The Long-Run Supply of New Reserves of Crude Oil in the United States, 1966-1973.** (Doctoral Dissertation, Vanderbilt University, 1975). 1979

Breed, Alice Gerster. **The Change in Social Welfare from Deregulation: The Case of the Natural Gas Industry.** (Doctoral Dissertation, Boston College, 1973). 1979

Carlson, Rodger D. **The Economics of Geothermal Power in California.** (Doctoral Dissertation, Claremont Graduate School, 1970). 1979

Challa, Krishna. **Investment and Returns in Exploration and the Impact on the Supply of Oil and Natural Gas Reserves.** (Doctoral Dissertation, Massachusetts Institute of Technology, 1974). 1979

Cookenboo, Leslie, Jr. **Crude Oil Pipe Lines and Competition in the Oil Industry** *and* **Costs of Operating Crude Oil Pipe Lines.** 1955/1954

Dale, Alfred George. **Nuclear Power Development in the United States to 1960.** (Doctoral Dissertation, University of Texas, 1975). 1979

Deegan, James F[lournoy]. **An Econometric Model of the Gulf Coast Oil and Gas Exploration Industry.** (Doctoral Dissertation, Southern Methodist University, 1975). 1979

Dillon, Robert John. **Reality and Value Judgment in Policymaking.** (Doctoral Dissertation, University of California, Los Angeles, 1974). 1979

DuBoff, Richard B. **Electric Power in American Manufacturing, 1889-1958.** (Doctoral Dissertation, University of Pennsylvania, 1964). 1979

Eichner, Donald O[scar]. **The Inter-American Nuclear Energy Commission.** (Doctoral Dissertation, The American University, 1969). 1979

Ellis, Theodore John. **Potential Role of Oil Shale in the U.S. Energy Mix.** (Doctoral Dissertation, Colorado State University, 1972). 1979

Erickson, Edward Walter. **Economic Incentives, Industrial Structure and the Supply of Crude Oil Discoveries in the U.S., 1946-58/59.** (Doctoral Dissertation, Vanderbilt University, 1968). 1979

Fenichel, Allen Howard. **Quantitative Analysis of the Growth and Diffusion of Steam Power in Manufacturing in the United States, 1838-1919.** (Doctoral Dissertation, University of Pennsylvania, 1964). 1979

Foster, Abram John. **The Coming of the Electrical Age to the United States.** (Doctoral Dissertation, University of Pittsburgh, 1952). 1979

Fulda, Michael. **Oil and International Relations.** (Doctoral Dissertation, The American University, 1970). 1979

Gessford, John Evans. **The Use of Reservoir Water for Hydroelectric Power Generation.** (Doctoral Dissertation, Stanford University, 1957). 1979

Gilbreth, Terry John. **Governing Geothermal Steam.** (Doctoral Dissertation, University of California, Riverside, 1974). 1979

Grayson, C. Jackson, Jr. **Decisions under Uncertainty: Drilling Decisions by Oil and Gas Operators.** 1960

Hall, Harry S. **Congressional Attitudes Toward Science and Scientists.** (Doctoral Dissertation, University of Chicago, 1961). 1979

Jacoby, Henry Donnan. **Analysis of Investment in Electric Power.** (Doctoral Thesis, Harvard University, 1967). 1979

Johnson, Charles J. **Coal Demand in the Electric Utility Industry, 1946-1990.** (Doctoral Thesis, Pennsylvania State University, 1972). 1979

Johnson, James P. **A "New Deal" for Soft Coal.** (Doctoral Dissertation, Columbia University, 1968). 1979

Keating, William Thomas. **Politics, Technology, and the Environment.** (Doctoral Dissertation, Indiana University, 1974). 1979

Kolb, Jeffrey Alan. **An Econometric Study of Discoveries of Natural Gas and Oil Reserves in the United States, 1948 to 1970.** (Doctoral Dissertation, University of Oregon, 1974). 1979

Lawrence, Anthony G. **Pricing and Planning in the U.S. Natural Gas Industry.** (Doctoral Dissertation, State University of New York at Buffalo, 1973). 1979

Lehman, Edward Richard. **Profits, Profitability, and the Oil Industry.** (Doctoral Dissertation, New York University, 1964). 1979

Manes, Rene Pierre. **The Effects of United States Oil Import Policy on the Petroleum Industry.** (Doctoral Thesis, Purdue University, 1961). 1979

Marcus, Kenneth Karl. **The National Government and the Natural Gas Industry, 1946-56.** (Doctoral Thesis, University of Illinois, 1962). 1979

McDonald, Philip R. **Factors Influencing Fuel Oil Growth.** (Doctoral Thesis, Harvard University, 1966). 1979

Meloe, Torleif. **United States Control of Petroleum Imports.** (Doctoral Dissertation, Columbia University, 1966). 1979

Nowill, Paul Henry. **Productivity and Technological Change in Electric Power Generating Plants.** (Doctoral Dissertation, University of Massachusetts, 1971). 1979

Pagoulatos, Angelos. **Major Determinants Affecting the Demand and Supply of Energy Resources.** (Doctoral Dissertation, Iowa State University, 1975). 1979

Pendergrass, Bonnie Baack. **Public Power, Politics, and Technology in the Eisenhower and Kennedy Years.** (Doctoral Dissertation, University of Washington, 1974). 1979

Phillips, David Gordon. **Federal-State Relations and the Control of Atomic Energy.** (Doctoral Dissertation, Syracuse University, 1964). 1979

Schramm, Gunter. **The Role of Low-Cost Power in Economic Development.** (Doctoral Dissertation, University of Michigan, 1967). 1979

Simon, Simon M. **Economic Legislation of Taxation.** (Doctoral Thesis, New York University, 1968). 1979

Smith, David Brian. **The Economics of Inter-Energy Competition in the United States.** (Doctoral Dissertation, University of Nebraska, 1971). 1979

Spann, Robert M. **The Supply of Natural Resources.** (Doctoral Thesis, North Carolina State University at Raleigh, 1970). 1979

Spooner, Robert Donald. **Response of Natural Gas and Crude Oil Exploration and Discovery to Economic Incentives.** (Doctoral Dissertation, University of Pennsylvania, 1973). 1979

Steele, Henry. **The Economic Potentialities of Synthetic Liquid Fuels from Oil Shale.** (Doctoral Dissertation, Massachusetts Institute of Technology, 1957). 1979

Striner, Herbert E. **An Analysis of the Bituminous Coal Industry in Terms of Total Energy Supply and a Synthetic Oil Program.** (Doctoral Dissertation, Syracuse University, 1951). 1979

Strout, Alan Mayne. **Technological Change and United States Energy Consumption, 1939-1954.** (Doctoral Dissertation, University of Chicago, 1967). 1979

Waltrip, John Richard. **Public Power During the Truman Administration.** (Doctoral Dissertation, University of Missouri, 1965). 1979

Wedemeyer, Karl Eric. **Interstate Natural Gas Supply and Intrastate Market Behavior.** (Doctoral Dissertation, University of Southern California, 1972). 1979

Whillier, Austin. **Solar Energy Collection and its Utilization for House Heating.** (Doctoral Thesis, Massachusetts Institute of Technology, 1953). 1979

Young, James Van. **Judges and Science: The Case Law on Atomic Energy.** (Doctoral Dissertation, University of Iowa, 1964). 1979

HD
9560.5
.F76
1979

HD
9560.5
.F76
1979